BRIDGE THE GAP MATHS™

the often-missing-but-really-need-to-know-bits

LAURIE BEESTING

Issue 1
Edited by: Laurie and Andy Beesting
Cover image by: Michelle Peters
Book and cover design by: Michelle Peters

ISBN: 978-1-9994385-1-7

Contents

Introduction

Welcome 7
Philosophy and About the Author 8
    Who am I and why have I created this resource? 8
    What is this resource? And what is it NOT? 10
    Who is it for? 10
Guidance for Use and Getting Started 11
    Principles and general outline for use...11
    Using the work / teaching pages 12
    Record keeping...13
    Order of activities and timings 14
    Resources..15
    Getting started 16
    Suggested first session & Next steps 17

Unit 1: Number

Unit 1 Introduction..19
Unit 1 'I CAN / I KNOW' Record Keeping Checklist 20
Four Rules of Number: Associated Vocabulary 21
Product and Difference..23
Worded Problems: How to Approach Them..25
Prefixes: Clues to Numbers and Amounts 27
Place Value for Whole Numbers 29
Approximating a Decimetre 31
Numbers of Days in the Months of the Year 33
Mean (Average) of a Set of Numbers 35
Mode of a Set of Numbers 37
Median of a Set of Numbers 39
15x Table: The First 5 Steps 41
Hiding Zero(s) to Make Multiplication Easier 43
Multiplying by One Digit: Column Multiplication 45
Multiplying by Two Digits: Column Multiplication..47

5s and 20s in One Hundred 49

Simple Doubling 51

Harder Doubling 53

Simple Halving 55

Harder Halving 57

Finding a Quarter of Numbers 59

Long Division 61

Short Division 63

Basic Measurement Facts 65

Unit 2: Fractions, Decimals and Percentages

Unit 2 Introduction 67

Unit 2 'I CAN / I KNOW' Record Keeping Checklist 68

Dividing into Thirds (then Sixths) 71

Dividing into Fifths (then Tenths) 73

Three Types of Fractions 75

Lowest Common Denominator (LCD) 77

Helpful Fractions Rap 79

Equivalent Fractions 81

Ordering or Comparing Fractions 83

Improper Fractions into Mixed Numbers (or Whole Numbers) .. 85

Improper Fractions: When to Change Them 87

Mixed Numbers into Improper Fractions 89

Simplifying Fractions: Chopping in Half 91

Simplifying Fractions: A Basic Rule 93

Simplifying Fractions: Times Tables Awareness 95

Simplifying Fractions: Choosing a Method 97

Adding or Subtracting Fractions with Same Denominators 99

Adding or Subtracting Fractions with Different Denominators .. 101

Multiplying Fractions 103

Dividing Fractions 105

Recognising Half: Alarm Bell Clues 107

Recognising Quarter: Alarm Bell Clues 109

Recognising Three-Quarters: Alarm Bell Clues 111

The 25x Table: The First 4 Steps 113

Fractions, Decimals and Percentages: Alarm Bell Clues 115

Simple Fractions of Numbers: Mentally 117

Harder Fractions of Numbers: Mentally 119
Place Value for Decimals 121
Tenths and Hundredths: Visualising and Understanding 123
Tenths and Hundredths: Equivalent Values 125
Ordering or Comparing Decimals 127
Rounding Decimals to Nearest Tenth (1 decimal place)...................................129
Rounding Decimals to Nearest Hundredth (2 decimal places) 131
Multiplying Decimals by 10, 100, 1000...133
Dividing Decimals by 10, 100, 1000 135
Changing Fractions into Decimals or Percentages 137
Changing Fractions into Decimals or Percentages Using a Calculator 139

Unit 3: Shape and Geometry

Unit 3 Introduction 141
Unit 3 'I CAN / I KNOW' Record Keeping Checklist 142
Parallel Lines 145
Intersecting Lines 147
Perpendicular Lines 149
The Term 'Congruent' 151
Understanding 'D' in 2D / 3D 153
2D shapes: Prefix Number Clues..155
2D Shapes: 'Regular' and 'Irregular' 157
Circle Vocabulary 159
Degrees in Circles 161
Types of Angles 163
Measuring Angles with a Protractor 165
Constructing Angles with a Protractor 167
Different Types of Triangles..169
Sketching Triangles 171
Accurate Construction of Triangles 173
Six Important Quadrilaterals 175
Sketching Quadrilaterals 177
Perimeter of 2D Shapes..179
Area of Squares and Rectangles 181
Area of Triangles 183
Transformations Vocabulary 185
3D Shapes: Names 187

3D Shapes: Properties Vocabulary...189
Prisms and Pyramids: Properties..191
Volume of Cubes and Cuboids..193

Unit 4: Times Tables, Exponents and Prime Numbers

Unit 4 Introduction..195
Unit 4 'I CAN / I KNOW' Record Keeping Checklist.......................................196
Getting Started..197
Every Visit Record Chart for Times Tables Progress.......................................198
The A, B, C, D Times Tables Model: What it is and How to use it....................199
Invaluable Teaching Resources / Ten Tips for Learning Times Tables...............200
The 2x Table..210
The 3x Table..212
The 4x Table..214
The 5x Table..216
The 6x Table..218
The 7x Table..220
The 8x Table..222
The 9x Table..224
The 10x Table..226
The 11x Table..228
The 12x Table..230
Square Numbers..233
Square Roots...235
Squaring Numbers: Exponent^2...237
Cubing Numbers: Exponent^3...239
Prime Numbers...241
Ultimate Challenge: All Times Tables Quick Facts...243

FLASH FACTS ☆™

FLASH FACTS☆™ Introduction..245
Guidance for Use...246
FLASH FACTS☆™ Questions and Answers...248

Introduction

Welcome to **Bridge the Gap Maths™**.
Together we can step in to make a difference.

Don't skip this information section... it's a 'must-read'.
Do make time to read these introductory pages *before* you start on the work pages; the information is **central to the success** of the work programme.

Philosophy and About the Author

Who am I and why have I created this resource?

I am Laurie Beesting. I qualified as a teacher with a Bachelor of Education Honours Degree in 1990 and taught in the British school system for over 20 years, before moving to Canada in 2011, where I began teaching maths privately, one-to-one.

My extensive teaching career has given me sound knowledge of how children learn best. I **know** that learning styles differ wildly, and I became a **patient and inventive maths teacher**, finding existing or developing my own creative, but always **simple** ways to get concepts across.

Whilst teaching maths one-to-one, I listed **common problem areas**; these became known as the **'OFTEN-MISSING-BUT-REALLY-NEED-TO-KNOW-BITS'**, and I found simple-as-possible ways to address them.

✱ *Most enquiries for 1:1 maths lessons were from parents with an 'emergency cry for help!'... their age 10,11,12 child (or sometimes older teen) was struggling – things had been ok up to about age 9, but then somewhere along the line things seemed to go wrong: they were 'all at sea' – seemingly now **lacking important basics**, losing confidence and generally feeling anxious and frustrated about their maths... **could I help?***

*Some of the students were in **mainstream school**, others were **home-schooled**.*

For ages 10,11,12,13+ or even adults: it was about **GAPS – the missing essentials**... the not-being-sure-what-bits to target and check up on. And about the frustration of **not knowing where to begin**; the heavy feeling in the stomach of knowing that going back through *all* the earlier curriculum was depressing and **actually not realistic or do-able**.

For age 9: sometimes a parent of an age 9 child wanted to **fast-track – to push ahead**. Although these enquiries were in the minority, I have used the programme with some younger students who had the knowledge and maturity to cope with next-level concepts; they LOVED the feeling of 'getting a little ahead of the game'.

✻ *Students loved the bite sized, simple approach – and it was working! Confidence was growing and skills were building. Students were animated and feeling more in control; it was wonderful to see self esteem rising. They LOVED watching the 'I CAN / I KNOW' statements build and took pride in ownership of being able to build the list of skills they now 'owned'.*

It was not obvious (and didn't matter) which 'earlier' Year/Grade the skill was 'from' – so older students kept their dignity and simply enjoyed the fact they had 'nailed it'. That felt so good... for them, and for me.

The more I thought about and planned work to help these individual students, the more I realised that what I was planning and delivering day in, day out, were the same things! Students were generally needing reinforcement of the **same** skills and concepts.

It dawned on me that what I was doing with my maths students could be shared with thousands more students **if I told parents what I do and how I do it!**

Thus, the idea of Bridge the Gap Maths™ was born.
I wanted to put 'me and the programme' into an easy to use book. So here I am.

✻ *When this resource was still in the making, one parent of a struggling age 11 student said to me, "I think of it (the approach / the book) as dealing with little pieces of maths litter that have been accidentally dropped along the maths journey, that now need sweeping up, collecting and sorting out – I need a copy, so let me know when it's ready."*

Author Laurie Beesting

What is this resource? And what is it NOT?

Bridge the Gap Maths™ EQUIPS and EMPOWERS **you the parent** to tutor your child, focussing in on **JUST the often-missing-essentials** from the Intermediate curriculum... to 'hit the ground running' in High School.

What started as a simple list of the skills, methods, tips and tricks I use in 1:1 lessons daily became **a script for you to use to help your child.** Some of these teaching methods and approaches differ slightly from those in mainstream resources. **It is me in a book** – almost word for word; it's what I do daily 1:1.

✱ *It has only **one page per concept**: it takes each learning objective and clearly states what it is, **in plain language**. It then tells you **what to say and do** to teach it – simply and effectively: **'WHAT to teach and HOW to teach it'.***

It is not attempting to cover a complete or exhaustive curriculum list for any Year or Grade.

It does not have every single maths concept you would see on a curriculum list for a particular Year/Grade, level or age group as there are already excellent maths books available which offer full coverage.

INSTEAD it has only the concepts I identified in my experience as being **common sticking points**, and this makes it **different**.

Who is it for?

It is for children, teens and adults.
It is mainly for children and adults to use together, but could also be used by older students or adults to **teach themselves**, or give them a chance to revisit some skills they missed, or didn't quite understand fully first time around.

✱ *Do not get too hung up or worry about ages, levels, Year groups or peer achievements. Instead, start at the student's current level, as it is the only place you can realistically start.*

If skills are missing, you could be age 12 or 99.

Guidance for Use and Getting Started

Principles and general outline for use

There are no set rules as different circumstances require different approaches. The amount of time available and reasons for using it will govern usage.

It might be used daily, weekly, twice weekly, at weekends or over a discrete period of time as a 'block project'. It could also be used in, e.g. school holidays.

Look at these 2 extremes:
- if to be used to **fill in 'emergency gaps'** for an older (e.g. age 11+) student, it might be used as an intense, daily programme over a period of weeks or months
- if to be used to **fast-track** a younger (e.g. age 9) student, it could be a summer project, or run alongside another maths scheme over a year or more as a supplement

✳ *Key principles*

Regular use: think through your time frame carefully
Repetition: keep coming back to skills until they are firmly in place
Revisiting: even after 'Achieving' a skill, after a while go back to it and check it is still hot – maybe have a session every 2 or 3 weeks devoted to revisiting 'old' skills
Variety: change the focus and activity style within the same session
Fun: have fun wherever possible, or at least a 'fun' (lighter) activity somewhere in the middle or towards the end of a session
Small bites and appropriate pace: 'go' with the student
Building up confidence: through revisiting the achieved 'I CAN / I KNOW' skills and loading on praise

Using the work / teaching pages

There are four main Units, plus **FLASH FACTS**☆™ :

- **Number**
- **Fractions, Decimals and Percentages**
- **Shape and Geometry**
- **Times Tables, Exponents and Prime Numbers**
- **FLASH FACTS**☆™: Fun, fast-response Q and A, based on concepts from all four Units

Each page follows the same **four-step** format for the student and 'helper' to work through methodically together:

Step 1
The learning objective: identified in the title, always starting with 'I CAN' or 'I KNOW'. Make a point of **slowly reading the title statement out loud** before starting on the content, so it is absolutely clear what the learning objective is; (after completing the page, ask the student if s/he can say what the 'I CAN' or 'I KNOW' skill was).

Step 2
The explanation: short and as simple as possible, but should be read slowly, carefully and thoroughly **without skipping sentences or taking short cuts**. Each detail is important; it may be necessary to work through and discuss it more than once before moving on.

Step 3
Eight 'Try' questions: the student and helper then work through **eight 'try' questions** which follow on the *same* page **using the explanation section as 'hold your hand' reference and guidance**. If revisiting a page, even though there are only eight questions, it is unusual that a student remembers them; doing the examples in random order is helpful. Do have the student make up two similar questions for the helper to solve, as making up appropriate questions shows sound understanding.

Step 4
Eight 'Test' questions: when the student understands and feels ready, s/he **turns over** to the *back* of the page to find **eight 'test' questions** to complete **without referring to the explanation, guidance or examples**. Answers are provided, but these should be covered with, e.g. a ruler or piece of card, and only revealed after answering either orally or on paper. It should not be viewed as a pass or fail 'test', but simply an opportunity to see if the 'I CAN / I KNOW' skill has sunk in; if it hasn't, then revisit another time when other skills are in place.

✱   *Unit 4, the Times Tables, Exponents and Prime Numbers Unit, has a slightly different format, but is still simple and methodical to use; **step by step guidance for use (a 'must-read')** is provided at the start. If Times Tables knowledge is very weak, it is a good idea to make an early start here.*

✱   *The final section is **FLASH FACTS** ☆™: a list of **rapid-response questions, answers and quick activities**, confirming the student's understanding of essential must-know facts; a randomly ordered collection of concepts, skills and tricks from Units 1 – 4. This section also has **'must-read' step by step guidance** at the start.*

Record keeping

'I CAN / I KNOW' checklists: a skills checklist is provided at the start of each Unit and should be used as a simple record keeping system to show **how many times** and **when** a skill is visited. There is a place to record 'Achieved' status, to record when the student and helper decide that the skill / concept is **fully in place**.

'Achieved' status: the checklists are helpful to refer to when the student is applying him/herself to other maths activities, e.g. school homework, as they can see that there is ***proof* that they *do have* the skill required** to solve a problem, as the 'I CAN / I KNOW' chart says so...

> *...the skill is in the head somewhere, and just needs **'fishing out'**.*
> *This encourages ownership and responsibility for learning.*

✱   *For maximum learning impact:*
  - *read out loud and discuss the **'I CAN / I KNOW' statement** before starting a work page*
  - *repeat before the 8 'try' questions*
  - *repeat before the 8 'test' questions*
  - *and finally, repeat once more at the end – then discuss progress / success*

  *This helps the student be **crystal clear about the learning objective**, tune in, focus and take **ownership** of the learning process. **The student him/herself** should tick the 'Achieved' box on the record chart.*

Also consider devising your own simple system to show when understanding of a skill is still weak, 'getting there' or good. An easy and quick system is to use **traffic light colours**: a small red circle in the corner of the page for 'weak', yellow for 'getting there' and green for 'good'. This serves as an instant visual sign to show which pages need more attention / revisiting.

Order of activities and timings

Don't worry about speed of covering concepts or pages; work at a pace that feels manageable and comfortable.

The four Units can be done in any order. However, if the student's Times Tables knowledge is weak, it is a good idea to make an early start on Unit 4. Times Tables should be practised regularly as they form a strong basis for other Units.

Within a Unit do aim to follow the order of pages, but a student could, for example, do some work from the Number Unit and some work from the Shape and Geometry Unit in the same sitting, to give a balance of harder number crunching and lighter, more fun visual work.

Consider doing a few minutes 'warm up' at the start of a study time, whether it be a Times Tables song, a counting activity (e.g. count in twenties from 680 to 900), a quick sketch of some shapes, a quick one question 'revisit' of a newly achieved skill etc. This will help get the student into gear. Times Tables quick facts or some **FLASH FACTS**☆™ may also be used as a warm up.

Bite sized amounts of time: work until the student is ready to stop or have a break, vary the choice of activities, choose an easier skill if there is little time and a harder skill if there is more time available. Try to have fun, don't worry if things don't immediately sink in – and keep it light.

Pace of work: work through each page at the student's pace. Even when 'Achieved' status is reached, revisit, repeat and consolidate from time to time. A fun rewards system is recommended.

For tougher skills pages: where I have found in my experience that the student needs to have focussed, extra high concentration levels, there is a fun 'warning': ⊕ so you can ensure that the timing is just right to address the few concepts that are maybe a little harder to grasp, e.g. column multiplication.

Revisit and repeat often: revisiting skills keeps them alive. Maybe use 'Achieved' skills as the warm up activity as things get going. This will also serve the purpose of feeling proud of maths knowledge already in place.

At an appropriate point, begin to use FLASH FACTS☆™ for fast recall and flipping around different skills. It is sometimes appropriate to have a full hour session using **FLASH FACTS**☆™ questions – see the 'guidance for use' pages at the beginning of the **FLASH FACTS**☆™ Unit.

Resources

Lots of paper: plain, squared and lined for neat work, and plenty of 'scrap paper'... also coloured paper, card and larger sheets (old wallpaper rolls?) are fun to work on. Whiteboards and dry wipe markers are fun to use sometimes and add variety.

Selection of work resources: pencils and sharpener, erasers, fun coloured pens, glue, sticky-notes, highlighters, paperclips... as many fun and colourful resources as possible.

Variety of maths equipment: rulers, calculator, protractor, set square triangle, tracing paper (or kitchen parchment), pair of compasses and scissors. Remember counters to use as 'counters', but also as reward tokens – different coloured ones worth different points (e.g. red 50 points, blue 20, white 10) drawn out of a bag are great fun and provide motivation; adding them up mentally at the end is an extremely valuable mental skill in itself.

Additional fun sticker charts: for good concentration or for not giving up if something is tricky to grasp. Wherever and whenever possible pour on praise as confidence building is key. I am always amazed at how important collecting stickers is to students – even older students and adults love them. I often write the reason for the sticker in small writing underneath as this gives them more significance.

Make the Times Tables stick as illustrated in Unit 4. Plan to do at least some Times Tables work every session. Even when all the Times Tables facts are fully known, turn them into division facts instead (e.g. 4 x 5 = 20 or 20 ÷ 4 = 5).

✱ *Use imaginative resources whenever possible: use counters / buttons to show steps of ten or hundred or to find half of amounts; use a calculator (a vital skill in itself) to self mark work; make up silly songs, raps, poems, sayings and rhymes to remember things; find online games or Times Tables songs; play darts to create a set of number scores to find the mean, mode and median; use real bottles to study litres and millilitres etc.*

Getting started

Read and familiarise yourself with how Units 1 – 3 work and how to use the 'I CAN / I KNOW' checklists. Make sure you have a full sized, opaque 30 cm ruler or, e.g. a thicker strip of card for some pages, to cover answers in the 'test' sections.

Read and digest how the Times Tables, Exponents and Prime Numbers Unit works: do read the 'guidance for use' pages and the information about Levels A, B, C and D to get a feel for where the student is.

Read and digest how FLASH FACTS☆™ works: do read the 'guidance for use' pages.

For answers and responses: consider making some activities purely oral, others practical and some written, and have a variety of all three within a session. Don't always be concerned about things being neat and beautifully presented.

Time needed? As a rough guide, once you're ready to go you will need approximately one hour for session one. This opening session will feel a little disjointed so just go with the flow. A first session is like this even with an experienced teacher – there is always some trial and error, finding what feels good, and then refining things.

Writing in the book? It is your choice whether or not to write in the book, but consider that you may wish to revisit, so have lots of scrap paper available on which to work. Answering orally is sometimes enough.

Keep sessions lively and varied: don't get stuck on one activity that has gone stale. Aim for small bites and different activities, e.g. after 20 minutes of long division, move onto something more practical. A weary student can quickly wake up if focus is redirected. Key word: **variety.**

Try to use / engage different parts of the brain by writing, drawing, sitting, standing, doing, checking on calculator, singing, laughing, celebrating small successes and making up questions for each other.

✱ *Remember:*
   *For maximum learning impact, make a point of regularly reading out the 'I CAN / I KNOW' statements to be aware at all times of **what the learning objective is.** The student him/herself should tick the 'Achieved' box on the 'I CAN / I KNOW' record chart when ready to do so.*

Suggested first session

Brain warm up for 5 minutes: maybe count up or back in number steps, draw all 2D or 3D shapes you can think of, ask random **+ – x ÷** questions whilst rolling a tennis ball back and forth, stretch out arms to estimate 20 cm or 100 cm, play 'quick response Times Tables facts', list some 4 sided shapes... anything you feel is age appropriate and fairly easy to get in the mood. Keep it lively and pacey.

Now 15/20 minutes on maybe the first couple of pages of the Number Unit, or the Fractions, Decimals and Percentages Unit. Be sure to read out 'I CAN / I KNOW' statements.

Discuss and evaluate Times Tables knowledge and confidence for 10/15 minutes to establish if Times Tables are pretty good, sort of ok... or terrible! If they are terrible, that's ok (you are not alone) and there is no need to worry as we have a plan! Note these observations to use as a starting point for launching the Times Tables, Exponents and Prime Numbers Unit. If Times Tables are very weak, make a note to put in some major work here.

Move on to something lighter for 15/20 minutes: maybe the Shape and Geometry Unit.

Maybe have a break and if age appropriate, award a sticker for concentrating so far.

Back to work for another 10/15 minutes depending on concentration levels – maybe let the student choose the Unit.

Tip: if concentration levels drop, rather than stop, try a change of activity – something active and practical can refresh and help refocus.

Next steps

Evaluate. Give praise. Discuss how it went, **together**. Encourage the student to take ownership of how you both use the programme; embrace the activities enjoyed the most.

The 'helper' then should take some time out to think about how it went generally, what direction to go in next and about where and how the record keeping / success levels should be kept (using the 'I CAN / I KNOW' checklists or another way as well if you prefer).

For planning notes for next steps maybe have a sheet of paper (divided up into say 8 boxes for 8 sessions to begin with) to make brief notes on things to revisit or move on to in future sessions. Do this whilst it is fresh in your mind (and do make notes mid session). Planning notes should be quick, brief and flexible. Always bear in mind that it is never good to spend ages on the same activity – variety is key; if something is going badly, then stop and move onto something else as it may work much better later when other concepts are in place.

For the next session, vary as appropriate; it will take a few sessions to get into the groove. Just take it slowly, one step at a time.

Every week or two, plan in a session devoted to revisiting 'old' skills. 'Achieved' skills do need revisiting, and as well as keeping them alive, this builds confidence – a central tool to success.

So all information delivered; you are ready to begin.

Enjoy... and good luck!

Laurie Beesting

Laurie Beesting

Work Unit 1: Number

This collection of concepts, skills and tricks is mainly based on the 'Four Rules of Number': Addition, Subtraction, Multiplication and Division.

It also deals with some other relevant number based concepts.

Some of the skills seem pretty straight forward at first glance. However, I have found in my experience that they are actually often missing, or at least not confidently in place. Think of them as a series of foundation building blocks.

Enjoy and celebrate the review process if the activity proves that the skill is already in place, and work on the ones that are perhaps not so solid.

For tougher skills pages, where I have found in my experience that the student needs to have focussed, ultra high concentration levels, there is a fun 'warning': (!) so you can ensure that the timing is just right to address the few concepts that are maybe a little harder to grasp, e.g. column multiplication.

For maximum learning impact and ownership of ability:
- Read out loud and discuss the 'I CAN / I KNOW' statement before starting a work page
- Repeat before the 8 'try' questions
- Repeat before the 8 'test' questions
- Finally, repeat once more at the end – then discuss whether you have 'Achieved' or need to revisit later

This helps the student be clear about the learning objective, tune in, focus and take ownership of the learning process. The student him/herself should tick the 'Achieved' box on the record chart.

Instructions for the 'pop-up-finger-method' can be found in the introduction to Unit 4.

The 'I CAN / I KNOW' Checklist for Unit 1: Number

	Page	I CAN / I KNOW	Tick & Date Visited	Achieved
	21	I KNOW the different words associated with $+ - x \div$		❏
	23	I KNOW what product and difference mean		❏
	25	I KNOW that if I see a worded problem, I can identify three golden rules		❏
	27	I KNOW that some prefixes (word openings) can give me a clue for a number or amount		❏
	29	I KNOW my whole number place values to the millions place		❏
	31	I CAN hold my hand to estimate a decimetre		❏
	33	I KNOW how many days are in every month by using the knuckles trick		❏
	35	I CAN find the mean (average) of a set of numbers		❏
	37	I CAN find the mode of a set of numbers		❏
	39	I CAN find the median of a set of numbers		❏
	41	I KNOW my 15x Table up to 75		❏
	43	I KNOW that if a number ends in zero(s), I can use a hiding trick to make multiplication easier		❏
!	45	I CAN multiply by one digit using column multiplication		❏
!	47	I CAN multiply by two digits using column multiplication		❏
	49	I KNOW there are five 20s in 100, and twenty 5s in 100		❏
	51	I CAN double simple numbers		❏
!	53	I CAN double harder numbers		❏
	55	I CAN find half of simple whole numbers		❏
!	57	I CAN find half of tricky whole numbers		❏
	59	I KNOW that to find a quarter of a number, I half it, then half it again		❏
!	61	I CAN do long division		❏
!	63	I CAN do short division		❏
	65	I KNOW basic number facts about length, capacity and weight		❏

I KNOW the different words associated with + − x ÷

OK! Let's check some basics! Look at these vocabulary lists for the four number operations: addition, subtraction, multiplication and division.

Check you know all the terms. Maybe copy the lists to make a large poster with fun illustrations and an example of a calculation to match each operation.

Challenge yourself to see how many you can list on paper without looking, or ask someone to say the words, and you say whether they are + − x ÷ .

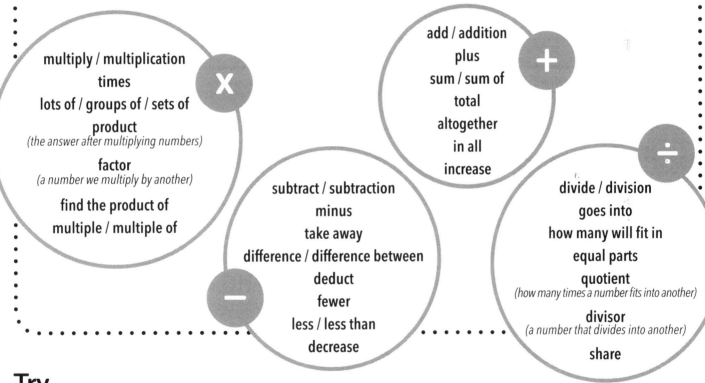

multiply / multiplication
times
lots of / groups of / sets of
product
(the answer after multiplying numbers)
factor
(a number we multiply by another)
find the product of
multiple / multiple of

X

add / addition
plus
sum / sum of
total
altogether
in all
increase

+

subtract / subtraction
minus
take away
difference / difference between
deduct
fewer
less / less than
decrease

−

divide / division
goes into
how many will fit in
equal parts
quotient
(how many times a number fits into another)
divisor
(a number that divides into another)
share

÷

Try

Work out these answers just to check you are familiar with the vocabulary used:

Total of 3 and 7

Name 2 factors to give a product of 12

Difference between 6 and 4

Sum of 10 and 5

Product of 8 and 5

4 less than 9

What is the quotient if I divide 6 by 2?

Product of 7 and 3

Now make up 2 more similar questions of your own!

Test

Work out these answers just to check you are familiar with the vocabulary used:

1.	Product of 5 and 3	5 x 3 = 15
2.	Sum of 5 and 3	5 + 3 = 8
3.	Difference between 5 and 3	5 – 3 = 2
4.	What is the quotient? $7\overline{)14}$	$7\overline{)14}^{\,2}$
5.	Find 2 factors to produce an answer (product) of 10	2 and 5 or 1 and 10
6.	Product of 10 and 3	10 x 3 = 30
7.	Find the difference between a tree measuring 7m high and a tree measuring 5m high	7 – 5 = 2 metres
8.	Total of 5, 3 and 1	5 + 3 + 1 = 9

Now re-read the title I KNOW statement; decide whether you have 'Achieved' or need to revisit.

I KNOW what product and difference mean

In my experience, for some reason a lot of people get caught out by these 2 terms. It just seems that they are the ones which fool us. Put a spotlight back on these again briefly, just to be 100% sure.

Don't let them catch YOU out!
Make a huge mental note of these 2 facts:

X **PRODUCT** times / multiply

− **DIFFERENCE** take-away / subtract

Try

Work out these answers to check that you are 100% sure of the terms 'product' and 'difference':

What is the product of 7 and 3?
Find the difference between 9 and 5
What is the difference between 50 and 40?
Find the product of 10 and 6
Find the difference between 14 and 2
The product of 5 and 5 is...?
What is the product of 2 and 3?
Find the difference between 21 and 20

Now make up 2 more similar questions of your own!

Test

Work out these answers to check that you are 100% sure of the terms 'product' and 'difference':

1. What is the difference between 18 and 8? $18 - 8 = 10$

2. What is the product of 2 and 9? $2 \times 9 = 18$

3. What is the difference between 6 and 4? $6 - 4 = 2$

4. The product of 7 and 3 is…? $7 \times 3 = 21$

5. The product of 4 and 4 is…? $4 \times 4 = 16$

6. Find the difference between 9 and 3 $9 - 3 = 6$

7. Find the product of 20 and 2 $20 \times 2 = 40$

8. What is the difference between 10 and 7? $10 - 7 = 3$

Now re-read the title I KNOW statement; decide whether you have 'Achieved' or need to revisit.

I KNOW that if I see a worded problem, I can identify three golden rules

If you see a worded problem, train your brain to say,

"Aha! A worded problem – I have to do 3 things:
HIGHLIGHT, SAUSAGE, RE-READ!"

1. **Highlight** the important information, starting with the numbers of course.

2. **Sausage**: write the ⟨ + − x ÷ ⟩ sausage on the page so you can decide which operation is needed, (and then **write out** your calculation 'sentence').

3. **Re-read** the question at the very end, after working it out. This step is REALLY important, as sometimes, we do some correct calculations but do not directly answer the original question.

e.g. Jo has a bookcase with 2 shelves. Each shelf holds 20 books which are all the same size. How many books will the bookcase hold? 2 x 20 = 40 (or 20 x 2 = 40)

The answer is: **40 books** (notice the answer is **underlined** for clarity)... then **RE-READ**!

Try

Solve these worded problems (you may find it helpful to draw pictures to visualise):

- Fred has 18 pencils and buys 12 more. How many does he now have?
- A farmer has 6 fields. Each field has 10 pigs. How many pigs are there altogether?
- Jack has 15 chickens and his dad buys him 11 more. How many chickens does Jack have now?
- There are 5 jars. Each jar holds 3 flowers. How many flowers are there altogether?
- Lila has 15 books. Her bookcase has 3 shelves. If she wants to put the same number of books on each shelf, how many books will go on each shelf?
- Leah has 15 flowers but 6 die. How many flowers are left?
- There are 29 cows in a field and 10 of them go indoors to be milked. How many cows are left in the field?
- A farmer has 18 sheep. He shares them equally between 3 fields. How many sheep go in each field?

Now make up 2 more similar questions of your own!

Test

Solve these worded problems (you may find it helpful to draw pictures to visualise):

1.	There are 4 baskets and 12 apples. If I put an equal number of apples into each basket, how many will be in each basket?	÷	$12 \div 4 = 3$ apples
2.	I have 3 boxes. Each box holds 12 pencils. How many pencils are there altogether?		$3 \times 12 = 36$ pencils
3.	I have 21 roses and 3 vases. If I share the roses out equally between the vases, how many will be in each?	÷	$21 \div 3 = 7$ roses
4.	Jack has 47 fish. He gets 13 more. Find the total number of fish he now has.	+	$47 + 13 = 60$ fish
5.	Sophie has 19 bracelets. She gives 13 to her friend. How many bracelets does Sophie now have?	-	$19 - 13 = 6$ bracelets
6.	I have 21 friends coming to my party. Then I invite 9 more. How many friends are now invited?	+	$21 + 9 = 30$ friends
7.	A farmer has 5 barns. Each barn holds 4 rolls of hay. How many rolls of hay are there altogether?		$5 \times 4 = 20$ rolls
8.	There are 40 Easter eggs in the store and 24 of them sell. How many eggs remain?		$40 - 24 = 16$ eggs

Now re-read the title I KNOW statement; decide whether you have 'Achieved' or need to revisit.

I KNOW that some prefixes (word openings) can give me a clue for a number or amount

Look at these prefixes and examples, and learn how prefixes can give us number clues:

dec:10
a **dec**agon is a 10 sided 2D shape

there are 10 **dec**imetres (dm) in a metre (m)

decimal means based on 10

a **dec**ahedron is a 10 faced 3D solid

a **dec**athlon is a sporting event with 10 activities

a **dec**ade is 10 years

cent:100
a **cent**ury is 100 years

there are 100 **cent**imetres (cm) in a metre (m)

a **cent**enarian is a 100 (or more) years old person

there are 100 **cent**s in a dollar

a **cent**enary is a 100 year anniversary

kilo:1,000
there are 1,000 metres (m) in a **kilo**metre (km)

a **kilo**watt (kW) is 1,000 watts (W) of power

there are 1,000 grams (g) in a **kilo**gram (kg)

Try

Look for 'prefix number clues' to solve these:

How many faces on a **dec**ahedron?

How many years in a **cent**ury?

How many **cent**s in a dollar?

How many g in a kg?

What do we call a 10 sided 2D shape?

A **cent**enary celebrates how many years?

How many m in a km?

How many sports did Fred do in the **dec**athlon?

Now make up 2 more similar questions of your own!

Test

Look for 'prefix number clues' to solve these:

1.	Mary is a centenarian. How old is she?	100 (or more) years old
2.	Which month was once the tenth month in the Roman calendar?	December
3.	How many sporting events in a decathlon?	10 events
4.	How many g in a kg?	1,000 g
5.	A 2D shape with 10 sides is called a...?	decagon
6.	How many m in 1 km?	1,000 m
7.	How many years in a century?	100 years
8.	How many decimetres fit into a metre?	10 dm

Now re-read the title I KNOW statement; decide whether you have 'Achieved' or need to revisit.

I KNOW my whole number place values to the millions place

Ok – grab a pencil and a ruler!
Copy this grid onto paper and try to memorise the box / place values:

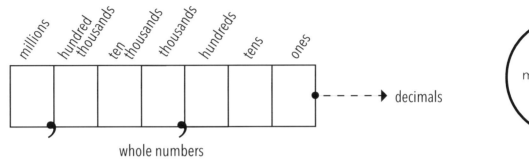

TIP
maybe colour code it!

Do this a few times until you can reproduce it quickly and confidently in less than a minute. Notice where the commas are placed: you count back 3 places from the right, then place a comma.

Create, write and say some big numbers which have either 4, 5, 6 or 7 digits. Say them out loud to get used to the wording, e.g. try saying **73,406**... the more you do, the better you will get.

Try

Say the whole numbers place value heading for the shaded box:

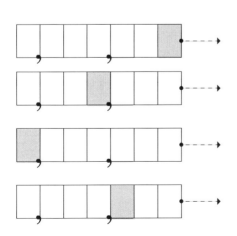

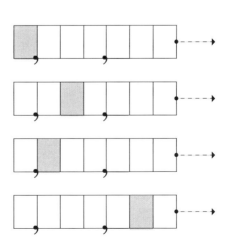

Test

Say the whole numbers place value heading for the shaded box:

1.		ones (1)
2.		tens (10)
3.		hundred thousands (100,000)
4.		ten thousands (10,000)
5.		hundreds (100)
6.		thousands (1,000)
7.		hundred thousands (100,000)
8.		millions (1,000,000)

Now re-read the title I KNOW statement; decide whether you have 'Achieved' or need to revisit.

I CAN hold my hand to estimate a decimetre

A decimetre (dm) is a distance of 10 cm and sometimes pops up in measurement work.

Say the word 'decimetre' out loud in a strong voice.

Make a C shape with your hand to see if you can make **exactly 10 cm** and see how close you are by using a ruler to check against. Try this a few times to see if you can get it exactly right.

There are 10 decimetres in a metre

1 dm = 10 cm

Try

Make a hand shape like the one above to estimate these distances.
Check against a ruler once you have guessed:

1 dm	Half a dm
0.25 (quarter) dm	One whole decimetre
$\frac{3}{4}$ dm	0.5 (half) dm
0.75 (three-quarters) of a dm	$\frac{1}{4}$ dm

Test

Make a hand shape to estimate these distances.
Check against a ruler once you have guessed:

1.   1 dm

2. $\frac{1}{2}$ dm

3.   75% (three-quarters) dm

4.   0.5 (half) dm

5.   100% (whole) dm

6.   half a dm

7.   0.25 (quarter) dm

8.   one whole dm

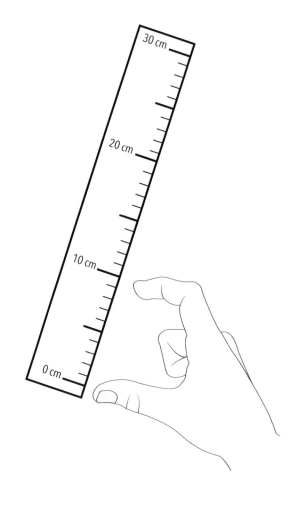

Now re-read the title I CAN statement; decide whether you have 'Achieved' or need to revisit.

I KNOW how many days are in every month by using the knuckles trick

 This trick is clever and is really easy to learn, but you DO need to know all the 12 months in order first.

Make your hands do this:

(It also actually works with only one hand if you go back to the first knuckle for August).

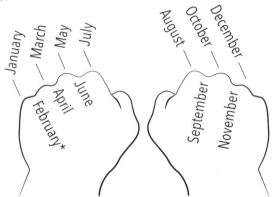

The **high** knuckles are **31 days** and the **low** dips are **30 days**.

*__February__ is an exception – it has 28 or 29 days depending on whether or not it falls in a leap year.

Try

Find the number of days in these months using the knuckles trick:

October

April

February

November

May

August

July

June

Now make up 2 more similar questions of your own!

Test

Find the number of days in these months using the knuckles trick:

1.	June	30 days
2.	September	30 days
3.	July	31 days
4.	January	31 days
5.	February	28 or 29 days
6.	August	31 days
7.	April	30 days
8.	December	31 days

Now re-read the title I KNOW statement; decide whether you have 'Achieved' or need to revisit.

I CAN find the mean (average) of a set of numbers

 There are two very simple steps to find the mean (average) of a set of numbers:

1. **Add them up (either mentally or using a calculator).**

2. **Divide the total by however many there are... DONE!**

Here are some pretend spelling scores – let's find the mean (average) score:

| 5 | 8 | 5 | 4 | 2 | 6 | 5 |

1. Add them up = 35

2. Divide the total by how many there are
    (there are 7 scores, so divide 35 by 7) = **5**

So the mean (average) is 5.

Try

Find the mean (average) of these sets of numbers (you may use a calculator):

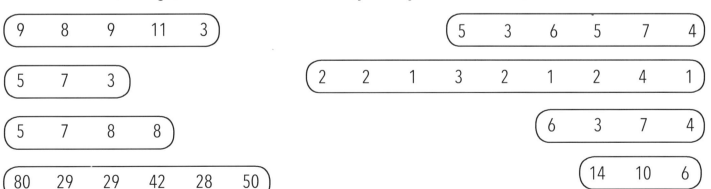

| 9 | 8 | 9 | 11 | 3 |

| 5 | 3 | 6 | 5 | 7 | 4 |

| 5 | 7 | 3 |

| 2 | 2 | 1 | 3 | 2 | 1 | 2 | 4 | 1 |

| 5 | 7 | 8 | 8 |

| 6 | 3 | 7 | 4 |

| 80 | 29 | 29 | 42 | 28 | 50 |

| 14 | 10 | 6 |

Now make up 2 more similar questions of your own!

Test

Find the mean (average) of these sets of numbers (you may use a calculator):

1.	What is the mean amount of flour used by a baker over a week? Mon 8 kg, Tues 4 kg, Wed 5 kg, Thurs 9 kg, Fri 5 kg, Sat 7 kg, Sun 11 kg	$8 + 4 + 5 + 9 + 5 + 7 + 11 = 49$ $49 \div 7 = 7$ kg
2.	Average spelling test score of these 7 children: 5, 8, 5, 4, 2, 6, 5	$5 + 8 + 5 + 4 + 2 + 6 + 5 = 35$ $35 \div 7 = 5$
3.	Mean darts score for 5 people: 9, 8, 9, 11, 13	$9 + 8 + 9 + 11 + 13 = 50$ $50 \div 5 = 10$
4.	Average number of glasses of water Fred had over 4 days: 7, 3, 5, 9	$7 + 3 + 5 + 9 = 24$ $24 \div 4 = 6$ glasses
5.	Mean weight: 5g, 7g, 8g, 8g	$5 + 7 + 8 + 8 = 28$ $28 \div 4 = 7$g
6.	What is the mean maths test score for Sally over 5 days of tests? Mon 17, Tues 18, Wed 16, Thurs 14, Fri 15	$17 + 18 + 16 + 14 + 15 = 80$ $80 \div 5 = 16$
7.	What is the mean age of these 5 children attending a dance club? Ages: 12, 10, 8, 10, 10 years	$12 + 10 + 8 + 10 + 10 = 50$ $50 \div 5 = 10$ years
8.	What is the mean number of apples sold over one week at the fruit store? 15, 12, 15, 3, 18, 6, 1	$15 + 12 + 15 + 3 + 18 + 6 + 1 = 70$ $70 \div 7 = 10$ apples

Now re-read the title I CAN statement; decide whether you have 'Achieved' or need to revisit.

I CAN find the mode of a set of numbers

 **Finding the mode of a set of numbers is just SO easy!
Look and say these 2 words out loud:**

They sound alike (or similar) because they both start with 'mo'!

Finding the mode is easy because it simply means the number which occurs the MO-ST times!

Look at these pretend spelling test scores: (7 8 7 4 2 6 7)

Which occurs the most times? The number **7** occurs the **most** times.

So the **mode is 7**. Done!

If there were 2 numbers that occurred 'most' times, we say the set of numbers is bimodal.

Try

Find the mode of these sets of numbers:

(7   7   5   6   3)

(4   3   4   3   1   2)

(3   4   3   9   3)

(3   6   6   9   6   12)

(5   2   7   5   6   4   5   6   5   3   5)

(8   2   8   1   2   8)

(12   10   4   6   12   11   12)

(1   2   1   3   1   4   1)

Now make up 2 more similar questions of your own!

Test

Find the mode of these sets of numbers:

1.	Number of cakes sold in a bakery over a week: 4, 0, 3, 4, 4, 9, 4	4 cakes
2.	Spelling test scores in Class B: 9, 10, 8, 5, 10, 4	10
3.	Number of children visiting the art gallery last week: 405, 390, 421, 293, 405, 500, 608	405 children
4.	Maths test results for class C: 14, 4, 20, 14, 19, 14, 9, 20, 19	14
5.	I threw seven darts – here were my scores: 17, 4, 8, 17, 15, 20, 3	17
6.	Number of children who attended art club over a week: 2, 3, 2, 6, 1, 2, 9	2 children
7.	Number of holidays Jack had over the last 4 years: 2, 1, 4, 2	2 holidays
8.	Times Tables test results in class G: 15, 16, 19, 20, 15, 19, 16, 15, 19	15 and 19 (bimodal)

Now re-read the title I CAN statement; decide whether you have 'Achieved' or need to revisit.

I CAN find the median of a set of numbers

**The word 'median' sounds a bit like 'medium'!
You know how 'medium' can suggest 'in the middle'?
Think about T shirt sizes: small, medium and large.**

The median is the number in the **middle** of the set of numbers, BUT the numbers have to be in **numerical order** first.

Let's take a set of pretend spelling test scores: ⟨ 5 8 5 4 2 6 5 ⟩

1. Put the numbers into numerical order: 2, 4, 5, **5**, 5, 6, 8

2. Now find the middle number... it is **5**

    So the **median is 5**.

(If there is an even number of spelling test scores, there will be 2 numbers in the middle... so add them both together and divide them by 2 to get the median.)

Try

Find the median of these sets of numbers:

⟨ 3   4   1   9   7 ⟩

⟨ 8   5   5   7   9   14   20 ⟩

⟨ 4   2   1   2   4   2   6 ⟩

⟨ 10   40   30 ⟩

⟨ 11   3   5   7 ⟩

⟨ 9   7   10   11   13 ⟩

⟨ 3   5   4   6   2 ⟩

⟨ 16   8   10   12 ⟩

Now make up 2 more similar questions of your own!

Test

Find the median of these sets of numbers:

1.	Spelling test scores: 5, 7, 4, 8, 10	4, 5, **7**, 8, 10
2.	Times Tables test results: 4, 6, 1, 2, 9, 12	1, 2, **4**, **6**, 9, 12 (so add 4 and 6 then divide by 2)... 5
3.	Ages of 5 children: 9, 10, 12, 7, 7 years	7, 7, **9**, 10, 12 years
4.	Dart scores: 2, 3, 14, 15, 7, 10, 12	2, 3, 7, **10**, 12, 14, 15
5.	The numbers 3, 10, 9, 15, 12, 17	3, 9, **10**, **12**, 15, 17 (add 10 and 12 then divide by 2)... 11
6.	Number of dogs: 8, 17, 9	8, **9**, 17 dogs
7.	Number of people at a stadium over three days: Mon 120, Tues 140, Wed 131	120, **131**, 140 people
8.	The numbers 1, 3, 5, 2, 10, 12	1, 2, **3**, **5**, 10, 12 (add 3 and 5 then divide by 2)... 4

Now re-read the title I CAN statement; decide whether you have 'Achieved' or need to revisit.

I KNOW my 15x Table up to 75

Oh good... this is a real quickie!
Rap, sing, shout, say, write it over and over, or make a poster...
but find a way to learn to count up in 15s until you hit 75!

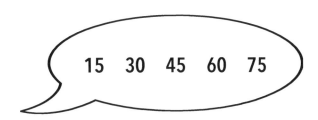

15   30   45   60   75

This is such an easy thing to learn and it is so, so useful – trust me!

Try

Use your 15x Table to solve these:

Count in 15s starting at zero: 0 _ _ _ _ 75

Count back in 15s from 75 to zero: 75 _ _ _ _ 0

15 + 15 + 15 = ?

5 x 15 = ?

How many 15s are there in 45?

What is half of 30?

What is 60 divided by 15?

One quarter of 60 is...?

Now make up 2 more similar questions of your own!

Test

Use your 15x Table to solve these:

1.	How many 15s are there in 30?	2
2.	Count up in 15s until you hit 75	15, 30, 45, 60, 75
3.	Multiply 15 by 3	45
4.	What are five fifteens?	75
5.	Write out the 15x Table (1 x 15 = ?, 2 x 15 = ? etc.) until you get to 5 x 15	1 x 15 = 15 2 x 15 = 30 3 x 15 = 45 4 x 15 = 60 5 x 15 = 75
6.	Start at 75, and count backwards in 15s until you reach zero	75, 60, 45, 30, 15, 0
7.	How many 15s are there in 60?	4
8.	What is 45 divided by 15?	3

Now re-read the title I KNOW statement; decide whether you have 'Achieved' or need to revisit.

I KNOW that if a number ends in zero(s), I can use a hiding trick to make multiplication easier

Look at this multiplication question: 5 x 40
Do you see 5 x 40... or 5 x 4?

See how you can hide the zero with a finger, so that it looks like 5 x 4 :

5 x 40   →   5 x 4 = 20

Then *reveal* the zero and adjust your answer to 20**0**!

Test the hiding trick with these calculations:

6**00** x 2   cover both zeros to work out 6 x 2,
              then reveal the zeros and adjust the answer... 12**00**

3**0** x 5   cover the zero to work out 3 x 5,
              then reveal the zero and adjust the answer... 15**0**

9**00** x 2   cover both zeros to work out 9 x 2,
              then reveal the zeros and adjust the answer... 18**00**

12**0** x 3   cover the zero to work out 12 x 3,
              then reveal the zero and adjust the answer... 36**0**

Try

Use the hiding zero(s) trick to solve these:

5 x 20	50 x 4	300 x 6	400 x 3
800 x 2	900 x 3	2 x 90	400 x 5

Now make up 2 more similar questions of your own!

Test

Use the hiding zero(s) trick to solve these:

1. 50 x 3   150

2. 800 x 2   1600

3. 800 x 5   4000

4. 6 x 60   360

5. 9 x 400   3600

6. 30 x 4   120

7. 120 x 5   600

8. 60 x 2   120

Now re-read the title I KNOW statement; decide whether you have 'Achieved' or need to revisit.

I CAN multiply by one digit using column multiplication

Oh no – the dreaded Times Tables!
This is how to multiply by one digit, laying out in columns:

Use grid paper to help you get the columns straight, or just be careful to **keep the place value positions lined up** if you use lined or plain paper. It is **vital** to keep the columns neat, ordered and straight.

Let's try out **576 x 4**.

1. Multiply the 4 by the top row number in the ones column: 4 x 6 = **24**
2. Place the 4 (from 24) in the ones column answer position and place the 2 (from 24) above the tens column (carry over 2)
3. Multiply the 4 by the top row number in the tens column: 4 x 7 = 28 then add on the 2 = **30**
4. Place the 0 (from 30) in the tens column answer position and place the 3 (from 30) above the hundreds column (carry over 3)
5. Multiply the 4 by the top row number in the hundreds column: 4 x 5 = 20 then add on the 3 = **23**
6. Place the 3 (from 23) in the hundreds column answer position and place the 2 (from 23) above the thousands column (carry over 2)
7. As there is no number in the top row thousands column we use 0: 4 x 0 = 0 then add on the 2 = **2**. Write this in the thousands column answer position.

```
    2  3  2
       5  7  6
  x          4
  _____
    2  3  0  4
```

Try

Make into column multiplications and then solve:

456 x 4	448 x 2	289 x 2	483 x 3
567 x 5	273 x 5	489 x 2	345 x 5

Now make up 2 more similar questions of your own!

Test

Make into column multiplications and then solve:

1. 438 x 2

```
    1
  4 3 8
x     2
-------
  8 7 6
```

2. 579 x 3

```
  1 2 2
  5 7 9
x     3
-------
1 7 3 7
```

3. 287 x 3

```
  2 2
  2 8 7
x     3
-------
  8 6 1
```

4. 458 x 5

```
  2 2 4
  4 5 8
x     5
-------
2 2 9 0
```

5. 299 x 5

```
  1 4 4
  2 9 9
x     5
-------
1 4 9 5
```

6. 373 x 4

```
  1 2 1
  3 7 3
x     4
-------
1 4 9 2
```

7. 992 x 6

```
  5 5 1
  9 9 2
x     6
-------
5 9 5 2
```

8. 455 x 4

```
  1 2 2
  4 5 5
x     4
-------
1 8 2 0
```

Now re-read the title I CAN statement; decide whether you have 'Achieved' or need to revisit.

I CAN multiply by two digits using column multiplication

Oh no – more Times Tables!
This is how to multiply by two digits, laying out in columns:

Use grid paper to help you get the columns straight, or just be careful to **keep the place value positions lined up** if you use lined or plain paper. It is **vital** to keep the columns neat, ordered and straight.

Let's try out **576 x 34**.

1. Write the calculation as a column multiplication and ignoring the '3' for now, multiply the top row of numbers by the 4 (as you did in one digit multiplication):

2. Erase the 'carry over' numbers (or cross them out).

3. Now we need to multiply the top row of numbers by the 3, **BUT... IT IS NOT 3... it is actually 30**, so you need to **place a zero** in the ones column answer position **before you multiply by the 3**. Ensure the 'carry overs' are placed above the same columns as before.

4. Finally, add the two answer rows together, using column addition... and then check your final answer on a calculator.

```
      2  3  2
      5  7  6
x        3  4
------------
      2  3  0  4
```

```
      1  2  1
      5  7  6
x        3  4
------------
      2  3  0  4
+  1  7  2  8  0
------------
   1  9  5  8  4
```

Try

Make into column multiplications and then solve:

456 x 32	287 x 23	387 x 32	452 x 54
448 x 53	235 x 46	245 x 44	883 x 25

Now make up 2 more similar questions of your own!

Test

Make into column multiplications and then solve:

1. 456 x 34

```
      4 5 6
x       3 4
    1 8 2 4
+ 1 3 6 8 0
  1 5 5 0 4
```

2. 234 x 35

```
      2 3 4
x       3 5
    1 1 7 0
+   7 0 2 0
    8 1 9 0
```

3. 443 x 25

```
      4 4 3
x       2 5
    2 2 1 5
+   8 8 6 0
  1 1 0 7 5
```

4. 247 x 45

```
      2 4 7
x       4 5
    1 2 3 5
+   9 8 8 0
  1 1 1 1 5
```

5. 345 x 36

```
      3 4 5
x       3 6
    2 0 7 0
+ 1 0 3 5 0
  1 2 4 2 0
```

6. 445 x 36

```
      4 4 5
x       3 6
    2 6 7 0
+ 1 3 3 5 0
  1 6 0 2 0
```

7. 289 x 42

```
      2 8 9
x       4 2
      5 7 8
+ 1 1 5 6 0
  1 2 1 3 8
```

8. 456 x 23

```
      4 5 6
x       2 3
    1 3 6 8
+   9 1 2 0
  1 0 4 8 8
```

Now re-read the title I CAN statement; decide whether you have 'Achieved' or need to revisit.

I KNOW there are five 20s in 100, and twenty 5s in 100

These two (easy to learn) facts pop up many times in maths questions – particularly in fractions, decimals and percentages work.

Say them out loud, in a strong voice a few times until you know them by heart:

There are five 20s in 100!

There are twenty 5s in 100!

You could make up a short, rhythmic rap to help you remember them.

Now, count up in fives until you hit 100. You should have counted **20** of them. Then, count up in twenties until you hit 100. You should have counted **5** of them.

Do this a few times to gain speed and confidence.

Try

Use the knowledge that there are five 20s in 100 and twenty 5s in 100 to solve these:

There are ____ fives in two hundred

$5\overline{)100}$

How many 20s are there in 300?

$20\overline{)200}$

How many times will twenty fit into four hundred?

Share 200 cows into 5 fields

Count up in multiples of five, until you hit one hundred – how many were there?

Count up in multiples of twenty, until you hit one hundred – how many were there?

Now make up 2 more similar questions of your own!

Test

Use your knowledge about 20s and 5s in 100 to solve these:

1.	How many fives fit into 200?	40
2.	$5\overline{)100}$	20
3.	How many groups of twenty fit into two hundred?	10
4.	$20\overline{)100}$	5
5.	Count up in multiples of twenty until you reach 300. How many twenties were there?	15
6.	There are ____ fives in 100	20
7.	There are ____ twenties in 100	5
8.	$200 \div 20$	10

Now re-read the title I KNOW statement; decide whether you have 'Achieved' or need to revisit.

I CAN double simple numbers

 Look at these simple doubling exercises – say the start number out loud and then DOUBLE IT mentally to reach the new number:

5 → 10   30 → 60   40 → 80   8 → 16   10 → 20   20 → 40

Fairly easy to do?

Now look at these further **doubling** examples and see if you can follow what is happening to the digits.

It is a clever method that can be applied to **doubling any number**, even 3, 4 or more digit numbers – it is called **partitioning** (splitting up) and **recombining** (putting back together by adding).

See how the **tens** double, the **ones** double... and then they **recombine**:

IMPORTANT
Remember that the tens column is TENS, e.g. in the number 23, the 2 is actually 20, not 2!

```
      23                    42                    34
     /  \                  /  \                  /  \
   40    6               80    4               60    8

  40 + 6 = 46          80 + 4 = 84          60 + 8 = 68
```

Try

Double these amounts mentally if you can, or on paper if it helps:

14 →   22 →   32 →   13 →

50 →   24 →   43 →   44 →

Now make up 2 more similar questions of your own!

Test

Double these amounts mentally if you can, or on paper if it helps:

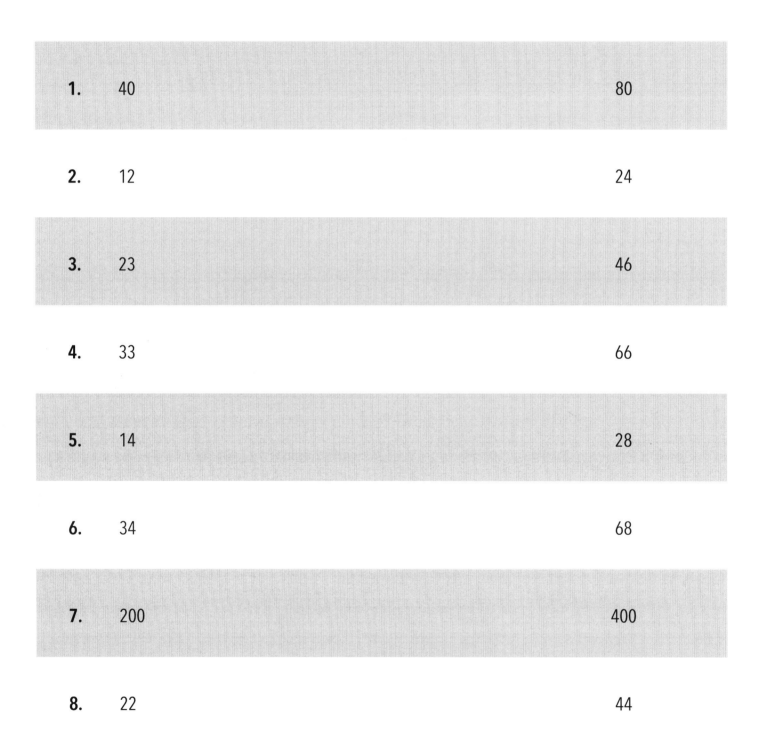

1. 40   80

2. 12   24

3. 23   46

4. 33   66

5. 14   28

6. 34   68

7. 200   400

8. 22   44

Now re-read the title I CAN statement; decide whether you have 'Achieved' or need to revisit.

I CAN double harder numbers

We can use partitioning to help double trickier numbers.

Look at these examples and see if you can follow what is happening to the digits.

It is a clever method that can be applied to **doubling any number**, even 3, 4 or more digit numbers – it is called **partitioning** (splitting up) and **recombining** (putting back together by adding).

See how the **hundreds** double, the **tens** double, the **ones** double... and then they **recombine**:

```
       36              72              68
      /  \            /  \            /  \
    60    12        140    4        120    16
```

60 + 12 = 72   140 + 4 = 144   120 + 16 = 136

> **IMPORTANT**
> Remember that the tens column is TENS, e.g. in the number 36, the 3 is actually 30, not 3, and the hundreds column is HUNDREDS, e.g. in the number 215, the 2 is actually 200 and the 1 is 10.

```
       48              17                215
      /  \            /  \             /  |  \
    80    16        20    14        400  20  10
```

80 + 16 = 96   20 + 14 = 34   400 + 20 + 10 = 430

Try

Double these amounts using the partitioning method:

48 →   35 →   39 →   47 →

53 →   126 →   85 →   139 →

Now make up 2 more similar questions of your own!

Test

Double these amounts using the partitioning method:

1.   48

```
        48
       /  \
     80    16
```

$80 + 16 = 96$

2.   45

```
        45
       /  \
     80    10
```

$80 + 10 = 90$

3.   29

```
        29
       /  \
     40    18
```

$40 + 18 = 58$

4.   146

```
          146
         / | \
      200  80  12
```

$200 + 80 + 12 = 292$

5.   57

```
        57
       /  \
    100    14
```

$100 + 14 = 114$

6.   38

```
        38
       /  \
     60    16
```

$60 + 16 = 76$

7.   108

```
         108
        / | \
     200  00  16
```

$200 + 00 + 16 = 216$

8.   82

```
        82
       /  \
    160    4
```

$160 + 4 = 164$

Now re-read the title I CAN statement; decide whether you have 'Achieved' or need to revisit.

I CAN find half of simple whole numbers

 Look at these simple halving exercises – say the start number out loud and then CHOP IT IN HALF mentally to get to the new number:

20 → 10   80 → 40   60 → 30   100 → 50   40 → 20   10 → 5

Fairly easy to do?

Now look at these further **halving** examples and see if you can follow what is happening to the digits.

It is a clever method that can be applied to **halving any number**, even 3, 4 or more digit numbers – it is called **partitioning** (splitting up) and **recombining** (putting back together by adding).

See how the **tens** halve, the **ones** halve... and then they **recombine**:

```
     24              48              66
    /  \            /  \            /  \
  10    2         20    4         30    3
```

10 + 2 = 12   20 + 4 = 24   30 + 3 = 33

IMPORTANT
Remember that the tens column is TENS, e.g. in the number 24, the 2 is actually 20, not 2!

Try

Find half of these numbers by chopping them in half mentally, or partitioning if you prefer:

80 →	82 →	26 →	64 →
60 →	46 →	88 →	28 →

Now make up 2 more similar questions of your own!

Test

Find half of these numbers by chopping them in half mentally, or partitioning if you prefer:

1. 68   34

2. 60   30

3. 84   42

4. 44   22

5. 80   40

6. 42   21

7. 12   6

8. 46   23

Now re-read the title I CAN statement; decide whether you have 'Achieved' or need to revisit.

I CAN find half of tricky whole numbers

 Finding half of trickier numbers can be challenging, but you should find this helpful. You need some counters.

Begin by thinking about these two statements:

- $\frac{1}{2}$ **of one (whole) is a half**:
  stop and think about what that means; maybe think pizza, a counter or an apple.

- $\frac{1}{2}$ **of 10 is 5**: that part is easy.

Lay out counters to work through these examples to find the half way point:

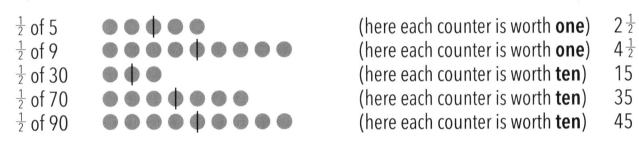

$\frac{1}{2}$ of 5	(here each counter is worth **one**)	$2\frac{1}{2}$
$\frac{1}{2}$ of 9	(here each counter is worth **one**)	$4\frac{1}{2}$
$\frac{1}{2}$ of 30	(here each counter is worth **ten**)	15
$\frac{1}{2}$ of 70	(here each counter is worth **ten**)	35
$\frac{1}{2}$ of 90	(here each counter is worth **ten**)	45

For these next examples, you will recognise **partitioning** – where we deal with tens and ones separately before **recombining**.

See how the tens halve, the ones halve, and then they recombine:

IMPORTANT
Remember that the tens column is TENS, e.g. in the number 94, the 9 is actually 90, not 9!

$$45 + 2 = 47 \qquad 5 + \tfrac{1}{2} = 5\tfrac{1}{2} \text{ or } 5.5 \qquad 15 + 1\tfrac{1}{2} = 16\tfrac{1}{2} \text{ or } 16.5$$

Try

Chop these numbers in half mentally, or use partitioning:

3 →	7 →	150 →	17 →
90 →	53 →	15 →	110 →

Now make up 2 more similar questions of your own!

Test

Chop these numbers in half mentally, or use partitioning:

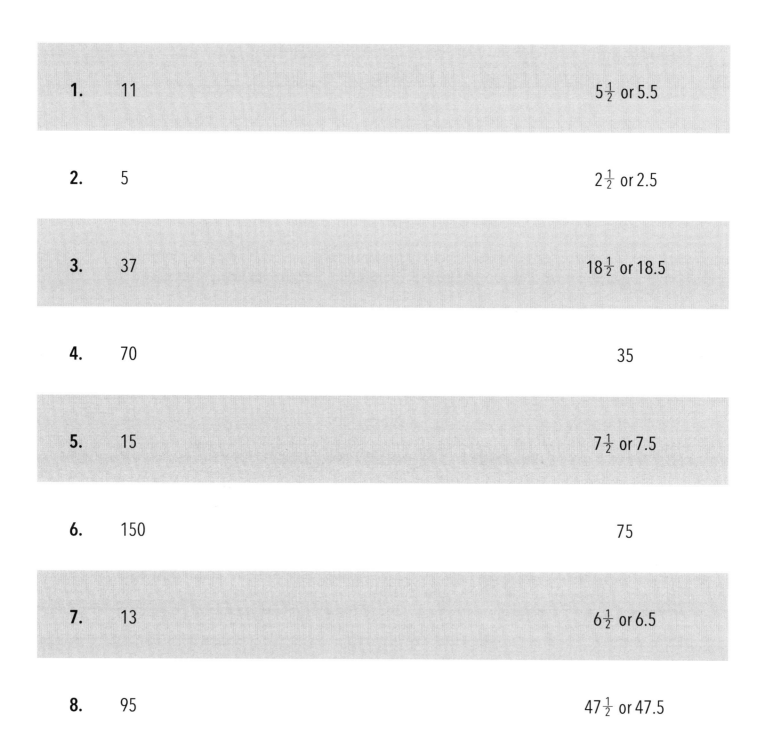

1.	11	$5\frac{1}{2}$ or 5.5
2.	5	$2\frac{1}{2}$ or 2.5
3.	37	$18\frac{1}{2}$ or 18.5
4.	70	35
5.	15	$7\frac{1}{2}$ or 7.5
6.	150	75
7.	13	$6\frac{1}{2}$ or 6.5
8.	95	$47\frac{1}{2}$ or 47.5

Now re-read the title I CAN statement; decide whether you have 'Achieved' or need to revisit.

I KNOW that to find a quarter of a number, I half it, then half it again

Finding a quarter ($\frac{1}{4}$) of a number is easy: you simply half it, then half it again.

Learn this simple saying:

> To find a quarter, I half it, then half it again!

Try it out with these numbers:

80	half it to **40**, then half it again to **20**
100	half it to **50**, then half it again to **25**
60	half it to **30**, then half it again to **15**
160	half it to **80**, then half it again to **40**
200	half it to **100**, then half it again to **50**

Try

Find a quarter of these numbers:

120 → 　　　　　800 → 　　　　　40 → 　　　　　16 →

160 → 　　　　　20 → 　　　　　400 → 　　　　　1,000 →

Now make up 2 more similar questions of your own!

Test

Find a quarter of these numbers:

1. 12   12 → 6 → 3

2. 80   80 → 40 → 20

3. 60   60 → 30 → 15

4. 120   120 → 60 → 30

5. 800   800 → 400 → 200

6. 16   16 → 8 → 4

7. 100   100 → 50 → 25

8. 140   140 → 70 → 35

Now re-read the title I KNOW statement; decide whether you have 'Achieved' or need to revisit.

I CAN do long division

Long division is one of the more complicated skills we need to address. You must concentrate hard on the series of steps involved.

Use grid paper to help you get the columns straight, or just be careful to **keep the place value positions lined up** if you use lined or plain paper. It is **vital** to keep the columns neat, ordered and straight.

1. Learn this rap. Say it over and over with a steady beat.

Write it like this in a sausage:

DIVIDE, TIMES, TAKE-AWAY, DROP!

$$\div$$
$$x$$
$$-$$
$$d$$

2. Using your non-writing hand, point to each operation in the sausage and do each in turn. **Work through the following example slowly and carefully** – one hand writing, and the other pointing to each operation and saying it out loud as you go.

Let's try out 786 ÷ 5. Point to each operation as you work through:

Step 1
how many 5s fit into 7? **1** (place the 1 above the 7)
1 times 5 = **5** (place the 5 below the 7)
7 take-away 5 = **2**
drop the **8** (so becomes 28)
now go back to the start of the sausage

Step 2
how many 5s fit in to 28? **5** (place the 5 above the 8)
5 times 5 = **25** (place the 25 below the 28)
28 take-away 25 = **3**
drop the **6** (so becomes 36)
now go back to the start of the sausage

Step 3
how many 5s fit into 36? **7** (place the 7 above the 6)
7 times 5 = **35** (place the 35 below the 36)
36 take-away 35 = **1**
there is nothing more to DROP, so remainder one (R1)

```
        1 5 7  R1
   5 ) 7 8 6
     -  5 ↓
        2 8
     -  2 5 ↓
          3 6
     -    3 5
            1
```

TIP
To find out how many times a number 'fits in', count up in the Times Table using the 'pop-up-finger-method'.

Try

Use the long division method. Lay these out on grid paper, and line up the columns carefully and neatly:

3)824   4)567   3)846   2)782

3)554   4)784   3)987   4)593

Now make up 2 more similar questions of your own!

Test

Use the long division method. Lay these out on grid paper, and line up the columns carefully and neatly:

1. $3\overline{)783}$

```
        2 6 1
    3 ) 7 8 3
      - 6
        1 8
      - 1 8
          0 3
      -     3
            0
```

2. $4\overline{)579}$

```
        1 4 4 R3
    4 ) 5 7 9
      - 4
        1 7
      - 1 6
          1 9
      -   1 6
            3
```

3. $3\overline{)678}$

```
        2 2 6
    3 ) 6 7 8
      - 6
        0 7
      -   6
          1 8
      -   1 8
            0
```

4. $5\overline{)984}$

```
        1 9 6 R4
    5 ) 9 8 4
      - 5
        4 8
      - 4 5
          3 4
      -   3 0
            4
```

5. $3\overline{)357}$

```
        1 1 9
    3 ) 3 5 7
      - 3
        0 5
      -   3
          2 7
      -   2 7
            0
```

6. $2\overline{)778}$

```
        3 8 9
    2 ) 7 7 8
      - 6
        1 7
      - 1 6
          1 8
      -   1 8
            0
```

7. $3\overline{)844}$

```
        2 8 1 R1
    3 ) 8 4 4
      - 6
        2 4
      - 2 4
          0 4
      -     3
            1
```

8. $5\overline{)679}$

```
        1 3 5 R4
    5 ) 6 7 9
      - 5
        1 7
      - 1 5
          2 9
      -   2 5
            4
```

Now re-read the title I CAN statement; decide whether you have 'Achieved' or need to revisit.

I CAN do short division

Most people prefer short division as it is faster than long division. It actually uses the same series of steps learned in long division, but takes short cuts.

The example here is the same question as you worked through for the long division example in the previous skill.

TIP
To find out how many times a number 'fits in', count up in the Times Table using the 'pop-up-finger-method'.

Let's try out 786 ÷ 5:

1. How many 5s fit into 7? **1** (place the 1 over the 7)
2. With how many left over? **2** (place a small 2 in front of the 8, so it looks like 28)
3. How many 5s fit into 28? **5** (place the 5 over the 28)
4. With how many left over? **3** (place a small 3 in front of the 6, so it looks like 36)
5. How many 5s fit into 36? **7** (place the 7 over the 36)
6. With how many left over? **1** (this is THE REMAINDER, so write R1)

$$5 \overline{)7\ ^28\ ^36} = 1\ \ 5\ \ 7\ R1$$

Try

Use the short division method:

$3 \overline{)575}$ $\qquad$ $5 \overline{)443}$ $\qquad$ $3 \overline{)567}$ $\qquad$ $5 \overline{)663}$

$4 \overline{)628}$ $\qquad$ $4 \overline{)321}$ $\qquad$ $6 \overline{)849}$ $\qquad$ $5 \overline{)246}$

Now make up 2 more similar questions of your own!

Test

Use the short division method:

1. $4\overline{)863}$

$$
\begin{array}{r}
2\ \ 1\ \ 5 \ \ \text{R3} \\
4\overline{)8\ \ 6\ \ ^23}
\end{array}
$$

2. $3\overline{)892}$

$$
\begin{array}{r}
2\ \ 9\ \ 7 \ \ \text{R1} \\
3\overline{)8\ \ ^29\ \ ^22}
\end{array}
$$

3. $4\overline{)567}$

$$
\begin{array}{r}
1\ \ 4\ \ 1 \ \ \text{R3} \\
4\overline{)5\ \ ^16\ \ 7}
\end{array}
$$

4. $5\overline{)265}$

$$
\begin{array}{r}
0\ \ 5\ \ 3 \\
5\overline{)2\ \ ^26\ \ ^15}
\end{array}
$$

5. $3\overline{)477}$

$$
\begin{array}{r}
1\ \ 5\ \ 9 \\
3\overline{)4\ \ ^17\ \ ^27}
\end{array}
$$

6. $6\overline{)843}$

$$
\begin{array}{r}
1\ \ 4\ \ 0 \ \ \text{R3} \\
6\overline{)8\ \ ^24\ \ 3}
\end{array}
$$

7. $4\overline{)327}$

$$
\begin{array}{r}
0\ \ 8\ \ 1 \ \ \text{R3} \\
4\overline{)3\ \ ^32\ \ 7}
\end{array}
$$

8. $3\overline{)889}$

$$
\begin{array}{r}
2\ \ 9\ \ 6 \ \ \text{R1} \\
3\overline{)8\ \ ^28\ \ ^19}
\end{array}
$$

Now re-read the title I CAN statement; decide whether you have 'Achieved' or need to revisit.

I KNOW basic number facts about length, capacity and weight

 We need to know some key measurement facts about length, capacity and weight.

Discuss generally what you already know about these measurement units:

- centimetres (cm), metres (m) and kilometres (km)
- millilitres (mL) and litres (L)
- grams (g) and kilograms (kg)

Look at these key measurement facts and see if you can spot the relationships between the numbers. Copy out the four statements to memorise:

For length / distance:

0 ⊤⊤⊤⊤⊤⊤⊤⊤⊤ 100 cm

1 m = 100 cm

For length / distance:

0 ▬ ▬ ▬ ▬ ▬ ▬ 1,000 m

1 km = 1,000 m

For capacity / liquid:

1 L = 1,000 mL

For weight / mass:

1 kg = 1,000 g

Now play around with some REAL resources:
- Look at **cm** and **m** on tape measures and measure the lengths of some items
- Look at **mL** and **L** number scales on liquid measuring containers and find references on milk containers, shampoo or water bottles etc.
- Look at **g** and **kg** on real weighing scales and weigh some items; find references on food tins and other groceries

Try

Fill in the missing parts to these statements about length, capacity and weight:

1 m = _____ cm   1,000 g = _____ kg   100 cm = _____ m   1,000 mL = _____ L

1 kg = _____ g   4 L = _____ mL   2 kg = _____ g   1 km = _____ m

Now make up 2 more similar questions of your own!

Test

Fill in the missing parts to these statements about length, capacity and weight:

1.   1 m = _____ cm   100 cm

2.   1,000 g = _____ kg   1 kg

3.   1 L = _____ mL   1,000 mL

4.   300 cm = _____ m   3 m

5.   1 kg = _____ g   1,000 g

6.   1 km = _____ m   1,000 m

7.   1,000 mL = _____ L   1 L

8.   2 kg = _____ g   2,000 g

Now re-read the title I KNOW statement; decide whether you have 'Achieved' or need to revisit.

Work Unit 2: Fractions, Decimals and Percentages

Fractions, decimals and percentages are all the same thing in a way – they all relate to parts of wholes, and are about dividing wholes into smaller parts. The whole can be a shape, a number or an amount.

Although they are often taught separately, it is important to understand that fractions, decimals and percentages are all closely related, and to see the links between them. This Unit encourages students to make these connections.

On a practical note, it is useful to have lots of circles (think 'pizzas') and rectangles (think 'chocolate bars') drawn on paper for this Unit, as FRACTION WORK is about chopping up wholes, and these simple visual references are very helpful.

 For tougher skills pages, where I have found in my experience that the student needs to have focussed, ultra high concentration levels, there is a fun 'warning': so you can ensure that the timing is just right to address the few concepts that are maybe a little harder to grasp, e.g. dividing fractions.

 For maximum learning impact and ownership of ability:
- Read out loud and discuss the 'I CAN / I KNOW' statement before starting a work page
- Repeat before the 8 'try' questions
- Repeat before the 8 'test' questions
- Finally, repeat once more at the end – then discuss whether you have 'Achieved' or need to revisit later

This helps the student be clear about the learning objective, tune in, focus and take ownership of the learning process. The student him/herself should tick the 'Achieved' box on the record chart.

 Instructions for the 'pop-up-finger-method' can be found in the introduction to Unit 4.

The 'I CAN / I KNOW' Checklist for Unit 2: Fractions, Decimals and Percentages

Page	I CAN / I KNOW	Tick & Date Visited	Achieved
71	I CAN roughly divide circles and rectangles into thirds (and therefore sixths)		❑
73	I CAN roughly divide circles and rectangles into fifths (and therefore tenths)		❑
75	I KNOW these 3 fraction types: proper, improper and mixed number		❑
(!) 77	I CAN find the lowest common denominator (LCD) for different fractions		❑
(!) 79	I KNOW the 'whatever you do to the top, you have to do to the bottom' rap for fractions		❑
81	I KNOW what equivalent fractions are and how to find them		❑
(!) 83	I CAN order or compare fractions by making them have the same denominator		❑
85	I KNOW how to change an improper fraction into a mixed number (or whole number)		❑
87	I KNOW when to change an improper fraction into a mixed number (or whole number)		❑
89	I CAN change a mixed number into an improper fraction		❑
91	I CAN help to simplify fractions by considering the 'chop in half' method first if both top and bottom are even numbers		❑
93	I CAN simplify fractions by finding a number that fits into both top and bottom		❑
95	I CAN simplify fractions by letting a Times Table jump out at me		❑
97	I CAN choose an appropriate simplifying fractions method		❑
99	I CAN add and subtract fractions with the same denominator		❑
(!) 101	I CAN add and subtract fractions with different denominators		❑
(!) 103	I CAN multiply fractions		❑
(!) 105	I CAN divide fractions		❑

Page	I CAN / I KNOW	Tick & Date Visited	Achieved
107	I CAN make an alarm bell go off in my brain for 'half' when I see 0.5 or 50%		❏
109	I CAN make an alarm bell go off in my brain for 'quarter' when I see 0.25 or 25%		❏
111	I CAN make an alarm bell go off in my brain for 'three-quarters' when I see 0.75 or 75%		❏
113	I KNOW my 25x Table to 100		❏
115	I CAN apply fractions, decimals and percentages alarm bells knowledge to real life maths problems		❏
117	I CAN find simple fractions of numbers mentally (where the numerator is one)		❏
⚠ 119	I CAN find harder fractions of numbers mentally (where the numerator is more than one)		❏
121	I KNOW where tenths and hundredths live in the decimal place value grid		❏
123	I KNOW that tenths are bigger than hundredths and can visualise them in a hundred square		❏
125	I KNOW that one tenth is the same as ten hundredths		❏
127	I CAN order or compare decimals by making them into fractions out of 100		❏
129	I CAN round decimals to the nearest tenth (one decimal place or 1 dp)		❏
131	I CAN round decimals to the nearest hundredth (two decimal places or 2 dp)		❏
133	I CAN multiply decimals by 10, 100 or 1000 by making the decimal point jump to the right		❏
135	I CAN divide decimals by 10, 100 or 1000 by making the decimal point jump to the left		❏
⚠ 137	I CAN change a fraction into a decimal or percentage by making it out of 100 (if the denominator fits exactly into 100)		❏
139	I CAN use a calculator to change a fraction into a decimal or percentage		❏

I CAN roughly divide circles and rectangles into thirds (and therefore sixths)

 Look at and copy / trace these fun drawings to divide circles and rectangles into thirds (three equal parts) and sixths (six equal parts):

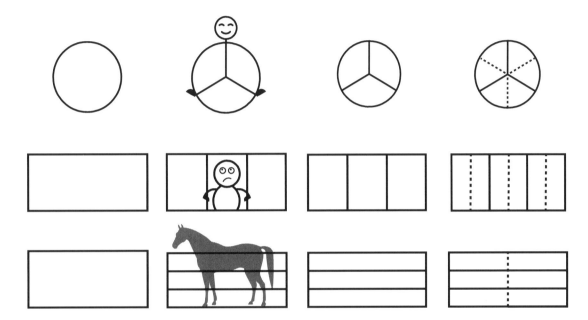

Thirds can be created easily – but **each third must be equal**, so play around until you get the lines in the right places.

To create the circles you can draw freehand or around a coin; for larger ones, draw around a small plate or make circles using a pair of compasses. Place a dot in the centre.

To create the rectangles you can draw freehand, use a ruler or find a template.

Remember to go onto make **sixths** by simply halving each **third**.

Try

Have fun: create lots of circles and rectangles on paper.

Divide them up into thirds but remember they should be equally divided into **three parts all the same size**. You will get better at it as you go. Once you have got the hang of making **thirds**, go back and make them all into **sixths**. Maybe colour them and make patterns.

Test

Follow these instructions:

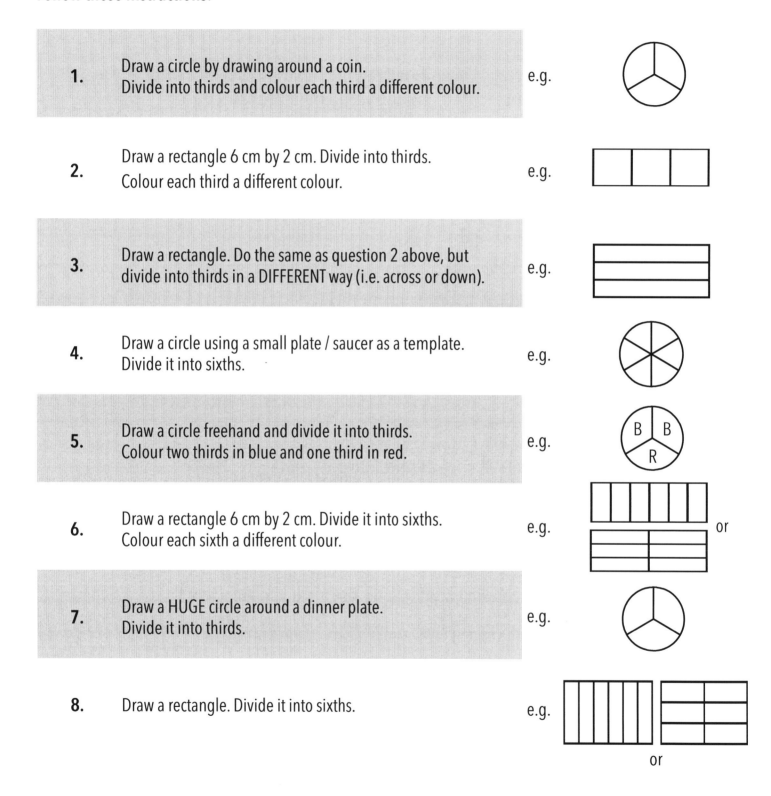

1. Draw a circle by drawing around a coin. Divide into thirds and colour each third a different colour.   e.g.

2. Draw a rectangle 6 cm by 2 cm. Divide into thirds. Colour each third a different colour.   e.g.

3. Draw a rectangle. Do the same as question 2 above, but divide into thirds in a DIFFERENT way (i.e. across or down).   e.g.

4. Draw a circle using a small plate / saucer as a template. Divide it into sixths.   e.g.

5. Draw a circle freehand and divide it into thirds. Colour two thirds in blue and one third in red.   e.g.

6. Draw a rectangle 6 cm by 2 cm. Divide it into sixths. Colour each sixth a different colour.   e.g.   or

7. Draw a HUGE circle around a dinner plate. Divide it into thirds.   e.g.

8. Draw a rectangle. Divide it into sixths.   e.g.   or

Now re-read the title I CAN statement; decide whether you have 'Achieved' or need to revisit.

I CAN roughly divide circles and rectangles into fifths (and therefore tenths)

 Look at and copy / trace these fun drawings to divide circles and rectangles into fifths (five equal parts) and tenths (ten equal parts):

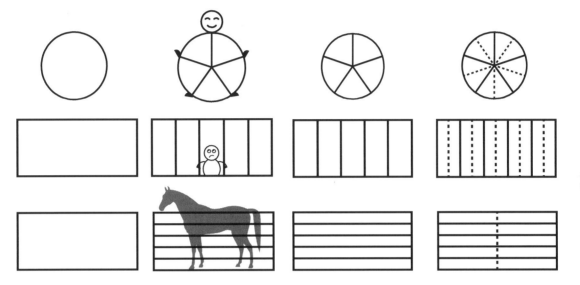

To create **fifths** in circles, draw a man with outstretched arms and legs. **Each fifth must be equal**, so play around until you get the lines in the right places. You will need to adjust how wide apart his legs and arms are to make the fifths equal. Stand and try it out with your own body.

To create the circles you can draw freehand or around a coin; for larger ones, draw around a small plate or make circles using a pair of compasses. Place a dot in the centre.

To create the rectangles you can draw freehand, use a ruler or find a template.

Remember to go onto make **tenths** by simply halving each **fifth**.

Try

Have fun: create lots of circles and rectangles on paper.

Divide them up into fifths but remember they should be equally divided into **five parts all the same size**. You will get better at it as you go. Once you have got the hang of making **fifths**, go back and make them all into **tenths**. Maybe colour them and make patterns.

Test

Follow these instructions:

1. Draw a circle by drawing around a coin.
   Divide into fifths and colour each fifth a different colour.

   e.g.

2. Draw a rectangle 10 cm by 5 cm. Divide it up into fifths.
   Colour each fifth a different colour.

   e.g.

3. Draw a rectangle 10 cm by 5 cm. Do the same as question 2 above, but divide into fifths in a DIFFERENT way (i.e. across or down).

   e.g.

4. Draw a circle using a small plate / saucer as a template.
   Divide it into tenths.

   e.g.

5. Draw a circle freehand and divide it into fifths. Colour two fifths in blue, one fifth in red and two fifths in green.

   e.g.

6. Draw a rectangle measuring 1 cm by 10 cm. Divide it into tenths. Colour each tenth a different colour.

   e.g.

7. Make a HUGE circle by drawing around a dinner plate.
   Divide it into fifths and then tenths.

   e.g.

8. Draw a rectangle measuring 20 cm by 10 cm.
   Divide it into tenths.

   e.g.

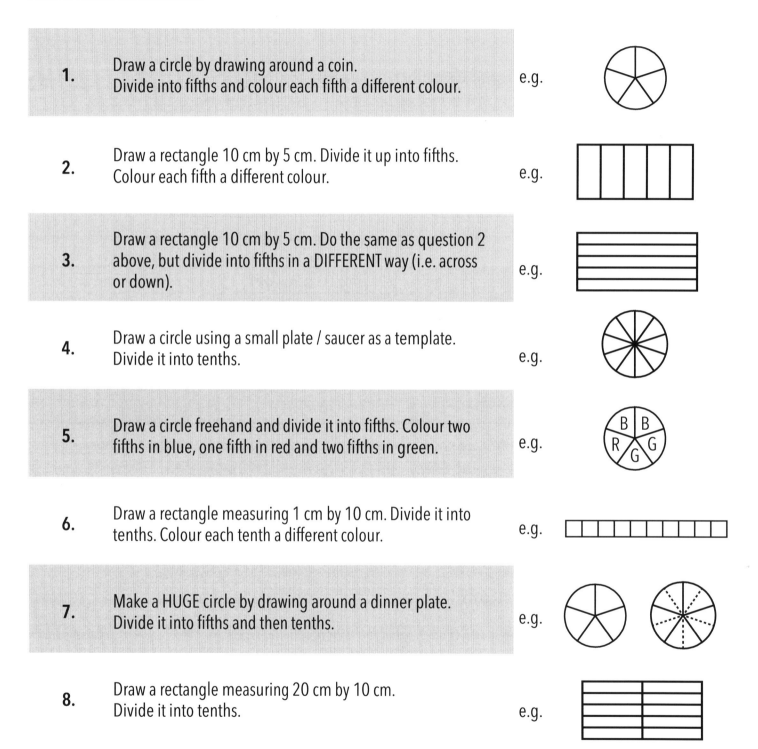

Now re-read the title I CAN statement; decide whether you have 'Achieved' or need to revisit.

I KNOW these 3 fraction types: proper, improper and mixed number

You need to be able to identify three TYPES of fraction: a proper fraction, an improper fraction and a mixed number.

A **proper fraction**: the top number (numerator) is smaller then the bottom number (denominator) e. g. $\frac{3}{4}$ $\frac{4}{5}$ $\frac{5}{12}$

An **improper fraction**: the top number (numerator) is bigger than or equal to the bottom number (denominator) e. g. $\frac{21}{7}$ $\frac{15}{6}$ $\frac{5}{5}$

A **mixed number**: has a whole number **and** a proper fraction

e. g. $3\frac{1}{4}$ $2\frac{1}{2}$ $7\frac{3}{4}$

Try

Identify these fractions as either proper, improper or mixed number:

$\frac{3}{5}$ $\frac{8}{2}$ $\frac{4}{7}$ $\frac{6}{10}$

$\frac{12}{8}$ $2\frac{1}{2}$ $\frac{15}{7}$ $1\frac{1}{2}$

Now make up 2 more similar questions of your own!

Test

Identify these fractions as either proper, improper or mixed number:

1. $3\frac{1}{2}$ mixed number

2. $\frac{17}{5}$ improper

3. $\frac{2}{3}$ proper

4. $\frac{4}{7}$ proper

5. $\frac{15}{15}$ improper

6. $6\frac{3}{4}$ mixed number

7. $8\frac{3}{4}$ mixed number

8. $\frac{16}{7}$ improper

Now re-read the title I KNOW statement; decide whether you have 'Achieved' or need to revisit.

I CAN find the lowest common denominator (LCD) for different fractions

 Sometimes we need to make different fractions have the same denominator. This is how to find the lowest common denominator (LCD).

Finding a common denominator for fractions enables us to compare their sizes, or maybe to add or subtract them.

Look at these 2 fractions: $\frac{1}{3}$ and $\frac{1}{4}$ They have **different** denominators.

To find the **lowest common denominator (LCD)**, find the **lowest number** that both denominators **will fit into**.

HOW: Count up in **3s** and then **4s** using the 'pop-up-finger-method': is there a number they will **both hit**? It is often enough just to do the first few steps of a Times Table as is the case here (writing them down helps visually):

3, 6, 9, (12), 15... (count up in the 3x Table)

4, 8, (12), 16... (count up in the 4x Table)

Aha! They both hit **12**! This denominator is the lowest one **common** to both, so 12 is the LCD for $\frac{1}{3}$ & $\frac{1}{4}$. The numerator has to adjust accordingly, but we will look at this later.

Another example:
Find the **lowest common denominator (LCD)** for $\frac{2}{5}$ and $\frac{3}{10}$.
Count up in both fives and tens and see which number both hit:

In fives: 5, (10) 15, 20, 25... In tens: (10) 20, 30, 40...

Therefore 10 is the lowest number both will fit into.
It is the **lowest common denominator (LCD)**.

TIP
I recommend using the 'pop-up-finger-method' and writing the steps of Times Tables out like this; it is good to count up and to see it visually.

Try

Find the lowest common denominator (LCD) for these pairs of fractions:

$\frac{1}{5}$ and $\frac{1}{2}$　　　$\frac{1}{5}$ and $\frac{1}{4}$　　　$\frac{2}{3}$ and $\frac{4}{6}$　　　$\frac{1}{5}$ and $\frac{1}{6}$

$\frac{1}{5}$ and $\frac{1}{3}$　　　$\frac{1}{8}$ and $\frac{1}{12}$　　　$\frac{1}{6}$ and $\frac{3}{10}$　　　$\frac{2}{3}$ and $\frac{1}{10}$

Now make up 2 more similar questions of your own!

Test

Find the lowest common denominator (LCD) for these pairs of fractions:

1. $\frac{1}{3}$ and $\frac{2}{4}$

    LCD is **12**
    3, 6, 9, (12) 15, 18...
    4, 8, (12) 16, 20, 24...

2. $\frac{3}{10}$ and $\frac{4}{5}$

    LCD is **10**
    (10) 20, 30, 40, 50, 60...
    5, (10) 15, 20, 25, 30...

3. $\frac{1}{6}$ and $\frac{1}{10}$

    LCD is **30**
    6, 12, 18, 24, (30) 36...
    10, 20, (30) 40, 50, 60...

4. $\frac{2}{3}$ and $\frac{2}{5}$

    LCD is **15**
    3, 6, 9, 12, (15) 18...
    5, 10, (15) 20, 25, 30...

5. $\frac{4}{5}$ and $\frac{3}{8}$

    LCD is **40**
    5, 10, 15, 20, 25, 30, 35, (40)..
    8, 16, 24, 32, (40) 48, 56, 64...

6. $\frac{1}{2}$ and $\frac{1}{7}$

    LCD is **14**
    2, 4, 6, 8, 10, 12, (14)..
    7, (14) 21, 28, 35, 42, 49...

7. $\frac{1}{2}$ and $\frac{2}{3}$

    LCD is **6**
    2, 4, (6) 8, 10, 12...
    3, (6) 9, 12, 15, 18...

8. $\frac{4}{5}$ and $\frac{1}{4}$

    LCD is **20**
    5, 10, 15, (20) 25, 30...
    4, 8, 12, 16, (20) 24...

Now re-read the title I CAN statement; decide whether you have 'Achieved' or need to revisit.

I KNOW the 'whatever you do to the top, you have to do to the bottom' rap for fractions

This is a useful rap to find equivalent fractions. Say it a few times to a steady rap beat – clap, tap or drum for the beat:

Look at what the rap does to these two example fractions:

> Whatever you do to the top, you have to do to the bottom... whatever you do to the bottom, you have to do to the top!

$\frac{5}{20}$

To change the bottom (denominator) to 100 (useful for making into decimals or percentages), we multiply it by 5, so we must also multiply the top (numerator) by 5:

$$\frac{5}{20}{}_{\times 5} \quad \rightarrow \quad \frac{5}{20}\begin{smallmatrix}\times 5\\\times 5\end{smallmatrix} \quad \rightarrow \quad \frac{25}{100}$$

Can you see how this has changed $\frac{5}{20}$ into $\frac{25}{100}$?
*(This is known as an **equivalent fraction** – see next skill).*

$\frac{6}{10}$

To change the top (numerator) to 3 (useful for simplifying), we divide it by 2, so we must also divide the bottom (denominator) by 2:

$$\frac{6}{10}{}^{\div 2} \quad \rightarrow \quad \frac{6}{10}\begin{smallmatrix}\div 2\\\div 2\end{smallmatrix} \quad \rightarrow \quad \frac{3}{5}$$

Can you see how this has changed $\frac{6}{10}$ into $\frac{3}{5}$?

TIP
We only **x** or **÷**
(never **+** or **-**)

Try

Work out what either the top or bottom has been multiplied or divided by and apply the rap to identify the missing digit:

$\frac{3}{20} = \frac{?}{100}$ $\qquad$ $\frac{6}{10} = \frac{?}{5}$ $\qquad$ $\frac{4}{8} = \frac{?}{40}$ $\qquad$ $\frac{3}{7} = \frac{30}{?}$

$\frac{14}{50} = \frac{7}{?}$ $\qquad$ $\frac{1}{2} = \frac{?}{8}$ $\qquad$ $\frac{3}{10} = \frac{30}{?}$ $\qquad$ $\frac{3}{4} = \frac{12}{?}$

Now make up 2 more similar questions of your own!

Test

Work out what either the top or bottom has been multiplied or divided by and apply the rap to identify the missing digit:

1. $\dfrac{4}{7} = \dfrac{?}{14}$ (x2) $\dfrac{4}{7} = \dfrac{\mathbf{8}}{14}$

2. $\dfrac{2}{5} = \dfrac{6}{?}$ (x3) $\dfrac{2}{5} = \dfrac{6}{\mathbf{15}}$

3. $\dfrac{12}{16} = \dfrac{6}{?}$ (÷2) $\dfrac{12}{16} = \dfrac{6}{\mathbf{8}}$

4. $\dfrac{5}{25} = \dfrac{?}{5}$ (÷5) $\dfrac{5}{25} = \dfrac{\mathbf{1}}{5}$

5. $\dfrac{3}{7} = \dfrac{30}{?}$ (x10) $\dfrac{3}{7} = \dfrac{30}{\mathbf{70}}$

6. $\dfrac{4}{20} = \dfrac{?}{100}$ (x5) $\dfrac{4}{20} = \dfrac{\mathbf{20}}{100}$

7. $\dfrac{16}{20} = \dfrac{?}{10}$ (÷2) $\dfrac{16}{20} = \dfrac{\mathbf{8}}{10}$

8. $\dfrac{3}{5} = \dfrac{?}{20}$ (x4) $\dfrac{3}{5} = \dfrac{\mathbf{12}}{20}$

Now re-read the title I KNOW statement; decide whether you have 'Achieved' or need to revisit.

I KNOW what equivalent fractions are and how to find them

 Equivalent fractions have the same 'value', but they look different.

Equivalent fractions take up the same amount of a whole but are simply split up into a different number of parts. Think about what these fractions below mean in terms of parts of a pizza or a chocolate bar. Look closely at the **numbers** in the written fractions and see how they relate to the pictures:

$\frac{3}{6}$ is equivalent to $\frac{1}{2}$

$\frac{2}{3}$ is equivalent to $\frac{4}{6}$

$\frac{1}{5}$ is equivalent to $\frac{2}{10}$

> Whatever you do to the top, you have to do to the bottom... whatever you do to the bottom, you have to do to the top!

To find an equivalent fraction:
- Simply **x** or **÷** top and bottom by the **same** number... any number!
- Maybe **x by 2, 3, 5, or 10**... or **÷ by 2 or 3** as these are quick and easy.

Try it out with the fraction $\frac{4}{6}$. This is how we could find some **equivalent fractions**:

x 2	multiply both top and bottom by 2	$\frac{4}{6}\,^{x2}_{x2}$ becomes	$\frac{8}{12}$
x 5	multiply both top and bottom by 5	$\frac{4}{6}\,^{x5}_{x5}$ becomes	$\frac{20}{30}$
÷ 2	divide both top and bottom by 2	$\frac{4}{6}\,^{÷2}_{÷2}$ becomes	$\frac{2}{3}$
x 10	multiply both top and bottom by 10	$\frac{4}{6}\,^{x10}_{x10}$ becomes	$\frac{40}{60}$

TIP
We only **x** or **÷**
(never **+** or **-**)

Try

Find equivalent fractions to:

$\frac{4}{8}$ $\frac{3}{4}$ $\frac{12}{20}$ $\frac{6}{8}$ $\frac{2}{9}$ $\frac{1}{4}$ $\frac{5}{6}$ $\frac{4}{10}$

Now make up 2 more similar questions of your own!

Test

Find equivalent fractions to:

1. $\frac{2}{8}$ e.g. $\frac{4}{16}$ $\frac{6}{24}$ $\frac{8}{32}$ $\frac{10}{40}$ $\frac{20}{80}$ $\frac{1}{4}$

2. $\frac{3}{10}$ e.g. $\frac{6}{20}$ $\frac{9}{30}$ $\frac{12}{40}$ $\frac{15}{50}$ $\frac{30}{100}$

3. $\frac{2}{5}$ e.g. $\frac{4}{10}$ $\frac{6}{15}$ $\frac{8}{20}$ $\frac{10}{25}$ $\frac{20}{50}$

4. $\frac{10}{30}$ e.g. $\frac{20}{60}$ $\frac{30}{90}$ $\frac{40}{120}$ $\frac{5}{15}$ $\frac{2}{6}$ $\frac{1}{3}$

5. $\frac{4}{6}$ e.g. $\frac{8}{12}$ $\frac{12}{18}$ $\frac{16}{24}$ $\frac{20}{30}$ $\frac{40}{60}$ $\frac{2}{3}$

6. $\frac{3}{9}$ e.g. $\frac{6}{18}$ $\frac{9}{27}$ $\frac{12}{36}$ $\frac{15}{45}$ $\frac{30}{90}$ $\frac{1}{3}$

7. $\frac{5}{15}$ e.g. $\frac{10}{30}$ $\frac{15}{45}$ $\frac{20}{60}$ $\frac{25}{75}$ $\frac{50}{150}$ $\frac{1}{3}$

8. $\frac{1}{3}$ e.g. $\frac{2}{6}$ $\frac{3}{9}$ $\frac{4}{12}$ $\frac{5}{15}$ $\frac{10}{30}$ $\frac{100}{300}$

Now re-read the title I KNOW statement; decide whether you have 'Achieved' or need to revisit.

I CAN order or compare fractions by making them have the same denominator

We often need to order or compare two or more fractions. It is about working out which fraction is biggest or smallest.

> Which fraction is smallest / biggest...

> Order the fractions from greatest to least or least to greatest...

> Put the fractions in order starting with the smallest / biggest...

> Order the fractions using < or >...

A method which ALWAYS works is to **make all the fractions have the same, common denominator** and then **adjust the numerators to make equivalent fractions**. Then you can easily see which is biggest / smallest.

Which is the bigger fraction here: $\frac{2}{6}$ or $\frac{4}{9}$? It is tempting to guess. **Do not guess.** Instead, **find the LCD** for **6** and **9** (the lowest number **both will fit into**), as explained earlier in this Unit.

HOW: First count up in sixes: 6, 12, ⟨18⟩ 24, 30...
Then count up in nines: 9, ⟨18⟩ 27, 36, 45...

The lowest number they both hit is 18. **Re-write** both fractions with denominator **18**.

Take each fraction in turn and **apply the rap** to adjust the numerators.

is equivalent to

$$\left(\frac{2}{6}\right) \rightarrow \frac{2^{\times 3}}{6_{\times 3}} = \left(\frac{6}{18}\right)$$

is equivalent to

$$\left(\frac{4}{9}\right) \rightarrow \frac{4^{\times 2}}{9_{\times 2}} = \left(\frac{8}{18}\right)$$

> Whatever you do to the top, you have to do to the bottom... whatever you do to the bottom, you have to do to the top!

Do you agree that spotting the bigger fraction out of $\frac{2}{6}$ and $\frac{4}{9}$ is much easier once the denominators are the same?

Hard to tell which is bigger: $\frac{2}{6}$ or $\frac{4}{9}$ So we make them have the same denominator:

Easy to tell which is bigger: $\frac{6}{18}$ or $\frac{8}{18}$ We can now easily see that $\frac{4}{9}$ is bigger than $\frac{2}{6}$.

Try

Compare these fractions by rewriting them with the LCD, and identify which is bigger:

$\frac{2}{3}$ and $\frac{2}{4}$ $\frac{4}{5}$ and $\frac{7}{10}$ $\frac{5}{8}$ and $\frac{3}{4}$ $\frac{2}{3}$ and $\frac{4}{5}$

$\frac{3}{10}$ and $\frac{2}{12}$ $\frac{9}{12}$ and $\frac{2}{3}$ $\frac{4}{5}$ and $\frac{3}{10}$ $\frac{1}{6}$ and $\frac{1}{5}$

Now make up 2 more similar questions of your own!

Test

Compare these fractions by rewriting them with the LCD, and identify which is bigger:

1. $\frac{3}{4}$ and $\frac{2}{3}$

   LCD = 12

   $\frac{3}{4} \rightarrow \frac{9}{12}$ and $\frac{2}{3} \rightarrow \frac{8}{12}$

   so $\frac{3}{4}$ is bigger

2. $\frac{6}{8}$ and $\frac{3}{5}$

   LCD = 40

   $\frac{6}{8} \rightarrow \frac{30}{40}$ and $\frac{3}{5} \rightarrow \frac{24}{40}$

   so $\frac{6}{8}$ is bigger

3. $\frac{8}{10}$ and $\frac{2}{3}$

   LCD = 30

   $\frac{8}{10} \rightarrow \frac{24}{30}$ and $\frac{2}{3} \rightarrow \frac{20}{30}$

   so $\frac{8}{10}$ is bigger

4. $\frac{4}{5}$ and $\frac{3}{6}$

   LCD = 30

   $\frac{4}{5} \rightarrow \frac{24}{30}$ and $\frac{3}{6} \rightarrow \frac{15}{30}$

   so $\frac{4}{5}$ is bigger

5. $\frac{1}{3}$ and $\frac{2}{4}$

   LCD = 12

   $\frac{1}{3} \rightarrow \frac{4}{12}$ and $\frac{2}{4} \rightarrow \frac{6}{12}$

   so $\frac{2}{4}$ is bigger

6. $\frac{2}{5}$ and $\frac{1}{2}$

   LCD = 10

   $\frac{2}{5} \rightarrow \frac{4}{10}$ and $\frac{1}{2} \rightarrow \frac{5}{10}$

   so $\frac{1}{2}$ is bigger

7. $\frac{2}{6}$ and $\frac{6}{9}$

   LCD = 18

   $\frac{2}{6} \rightarrow \frac{6}{18}$ and $\frac{6}{9} \rightarrow \frac{12}{18}$

   so $\frac{6}{9}$ is bigger

8. $\frac{4}{5}$ and $\frac{3}{4}$

   LCD = 20

   $\frac{4}{5} \rightarrow \frac{16}{20}$ and $\frac{3}{4} \rightarrow \frac{15}{20}$

   so $\frac{4}{5}$ is bigger

Now re-read the title I CAN statement; decide whether you have 'Achieved' or need to revisit.

I KNOW how to change an improper fraction into a mixed number (or whole number)

We need to learn the 'trick' of changing an improper fraction into a mixed number (or just a whole number): divide the top (numerator) by the bottom (denominator).

Look at this fraction: $\frac{15}{2}$ It is an **improper fraction**; it is top heavy.

To simplify, **divide the top by the bottom**:
If it doesn't fit in exactly, you will have a **mixed number**;
if it fits in exactly, you will have a **whole number**.

> Say, "How many times can you fit the bottom number into the top?"

- How many times does **2** fit into **15**? (count up in 2s using the 'pop-up-finger-method')
- It fits in **7** times, therefore there are **7 whole ones**.
- How many left over? ONE. **The denominator does not change**, so it is one over two $\frac{1}{2}$

Therefore, $\frac{15}{2}$ changed to a mixed number is $7\frac{1}{2}$

More examples:

$\frac{17}{5}$:   How many 5s fit into 17?   3 with $\frac{2}{5}$ left over: $\mathbf{3\frac{2}{5}}$

$\frac{25}{3}$:   How many 3s fit into 25?   8 with $\frac{1}{3}$ left over: $\mathbf{8\frac{1}{3}}$

$\frac{20}{5}$:   How many 5s fit into 20?   Exactly 4 fit in: **4**

> **TIP**
> Divide the numerator (top) by the denominator (bottom). The denominator stays the **same**.

Try

Change these improper fractions into mixed numbers (or whole numbers):

$\frac{17}{3}$ $\frac{12}{5}$ $\frac{14}{3}$ $\frac{13}{7}$ $\frac{14}{2}$ $\frac{16}{5}$ $\frac{18}{2}$ $\frac{9}{4}$

Now make up 2 more similar questions of your own!

Test

Change these improper fractions into mixed numbers (or whole numbers):

1. $\dfrac{16}{3}$ $5\frac{1}{3}$

2. $\dfrac{15}{2}$ $7\frac{1}{2}$

3. $\dfrac{12}{4}$ 3

4. $\dfrac{18}{5}$ $3\frac{3}{5}$

5. $\dfrac{20}{3}$ $6\frac{2}{3}$

6. $\dfrac{20}{2}$ 10

7. $\dfrac{17}{4}$ $4\frac{1}{4}$

8. $\dfrac{19}{5}$ $3\frac{4}{5}$

Now re-read the title I KNOW statement; decide whether you have 'Achieved' or need to revisit.

I KNOW when to change an improper fraction into a mixed number (or whole number)

Look at this improper fraction. Listen to what he sometimes shouts out:

$$\frac{19}{3}$$

"HEY! Don't leave me improper! Change me into a mixed or whole number! I wanna be a mixed or whole number!"

Improper fractions often need to be changed into **mixed numbers** (or just a whole number). This is called **simplifying** or putting a fraction into its simplest form.

If you do a fraction question and your **answer** ends up being an improper fraction, try to spot it and automatically go on to simplify.

Look at this example of adding 2 fractions: $\frac{4}{5} + \frac{3}{5} = \frac{7}{5}$

- The answer is an **improper** fraction. **Listen to what he is shouting out!**
- It should be changed to a mixed number straight away.
  Say, "How many 5s fit into 7? ONE with $\frac{2}{5}$ left over": $1\frac{2}{5}$

Your target is for you to **notice** whenever an answer is an improper fraction and put it into its simplest form, without you needing to have it pointed out.

Try

Read these fraction additions out loud. Simplify the answers if necessary:

$$\frac{8}{10} + \frac{3}{10} = \frac{11}{10} \qquad \frac{4}{5} + \frac{1}{5} = \frac{5}{5} \qquad \frac{4}{8} + \frac{6}{8} = \frac{10}{8} \qquad \frac{2}{4} + \frac{3}{4} = \frac{5}{4}$$

$$\frac{3}{4} + \frac{3}{4} = \frac{6}{4} \qquad \frac{7}{8} + \frac{2}{8} = \frac{9}{8} \qquad \frac{5}{9} + \frac{3}{9} = \frac{8}{9} \qquad \frac{3}{5} + \frac{1}{5} = \frac{4}{5}$$

Now make up 2 more similar questions of your own!

Test

Read these fraction additions out loud. Simplify the answers if necessary:

1. $\frac{9}{10} + \frac{2}{10} = \frac{11}{10}$ $\frac{11}{10} \rightarrow 1\frac{1}{10}$

2. $\frac{5}{7} + \frac{5}{7} = \frac{10}{7}$ $\frac{10}{7} \rightarrow 1\frac{3}{7}$

3. $\frac{3}{8} + \frac{6}{8} = \frac{9}{8}$ $\frac{9}{8} \rightarrow 1\frac{1}{8}$

4. $\frac{5}{9} + \frac{2}{9} = \frac{7}{9}$ $\frac{7}{9}$ (proper fraction)

5. $\frac{2}{3} + \frac{2}{3} = \frac{4}{3}$ $\frac{4}{3} \rightarrow 1\frac{1}{3}$

6. $\frac{5}{8} + \frac{6}{8} = \frac{11}{8}$ $\frac{11}{8} \rightarrow 1\frac{3}{8}$

7. $\frac{4}{10} + \frac{9}{10} = \frac{13}{10}$ $\frac{13}{10} \rightarrow 1\frac{3}{10}$

8. $\frac{5}{7} + \frac{1}{7} = \frac{6}{7}$ $\frac{6}{7}$ (proper fraction)

Now re-read the title I KNOW statement; decide whether you have 'Achieved' or need to revisit.

I CAN change a mixed number into an improper fraction

We need to learn the 'trick' of changing a mixed number into an improper fraction. It is usually the other way round, but this reverse skill is needed sometimes.

To change a **mixed number** into an **improper fraction**, **multiply** the **whole number** by the **denominator**, then **add on the numerator**.

Look at this fraction: $3\frac{4}{5}$ It is a mixed number: 3 whole ones and 4 fifths
(think 3 whole pizzas and $\frac{4}{5}$ of a pizza)

To see how many fifths there are in $3\frac{4}{5}$, change it into an improper fraction:
multiply the **whole number** by the **denominator**, then **add on the numerator**.

3 times 5 is 15 (there are 15 fifths in **3** whole ones – picture 3 whole pizzas in your head), plus the other $\frac{4}{5}$, will give 19 fifths: $\frac{19}{5}$. Notice that the denominator **does not** change.

More examples:

$6\frac{1}{2}$: 6 x 2 = 12 ➔ 12 + 1 (the numerator) = 13 ➔ $\frac{13}{2}$ (the denominator stays the same)

$3\frac{1}{2}$: 3 x 2 = 6 ➔ 6 + 1 (the numerator) = 7 ➔ $\frac{7}{2}$ (the denominator stays the same)

TIP
Multiply the whole number by the denominator, then add on the numerator. The denominator stays the same.

Try

Change these mixed numbers into improper fractions:

$2\frac{1}{6}$ $3\frac{2}{3}$ $5\frac{2}{5}$ $1\frac{5}{6}$

$4\frac{1}{4}$ $1\frac{1}{2}$ $2\frac{7}{8}$ $4\frac{3}{4}$

Now make up 2 more similar questions of your own!

Test

Change these mixed numbers into improper fractions:

1. $3\frac{2}{3}$ $\frac{11}{3}$

2. $4\frac{2}{3}$ $\frac{14}{3}$

3. $1\frac{5}{8}$ $\frac{13}{8}$

4. $2\frac{2}{5}$ $\frac{12}{5}$

5. $4\frac{1}{3}$ $\frac{13}{3}$

6. $1\frac{2}{9}$ $\frac{11}{9}$

7. $2\frac{3}{4}$ $\frac{11}{4}$

8. $3\frac{1}{8}$ $\frac{25}{8}$

Now re-read the title I CAN statement; decide whether you have 'Achieved' or need to revisit.

I CAN help to simplify fractions by considering the 'chop in half' method first if both top and bottom are even numbers

Simplifying a fraction means putting it into its lowest equivalent terms.

If both the top and bottom of a fraction are even numbers, you can chop them in half to get smaller numbers to make an equivalent fraction in lower terms.

I like to say:

"Oh! It is 'CHOP-IN-HALF-ABLE'!"

Then I chop both the top and the bottom numbers in half.
Sometimes it takes more than one 'chop in half' to get to lowest terms.

Look at these fractions. They are all 'CHOP-IN-HALF-ABLE':

$$\frac{6}{8} \rightarrow \frac{3}{4}$$

$$\frac{8}{12} \rightarrow \frac{4}{6} \rightarrow \frac{2}{3}$$

$$\frac{20}{24} \rightarrow \frac{10}{12} \rightarrow \frac{5}{6}$$

> **TIP**
> The 'chop in half method' will sometimes leave you with a fraction that can be simplified further, but at least you will have numbers that are easier to work with.

Try

Simplify these fractions using the 'chop in half' method:

$$\frac{4}{8} \qquad \frac{12}{20} \qquad \frac{2}{14} \qquad \frac{12}{16} \qquad \frac{4}{12} \qquad \frac{2}{10} \qquad \frac{20}{32} \qquad \frac{4}{10}$$

Now make up 2 more similar questions of your own!

Test

Simplify these fractions using the 'chop in half' method:

1. $\dfrac{12}{20}$ $\dfrac{6}{10} \rightarrow \dfrac{3}{5}$

2. $\dfrac{2}{4}$ $\dfrac{1}{2}$

3. $\dfrac{10}{18}$ $\dfrac{5}{9}$

4. $\dfrac{8}{12}$ $\dfrac{4}{6} \rightarrow \dfrac{2}{3}$

5. $\dfrac{2}{24}$ $\dfrac{1}{12}$

6. $\dfrac{20}{32}$ $\dfrac{10}{16} \rightarrow \dfrac{5}{8}$

7. $\dfrac{4}{14}$ $\dfrac{2}{7}$

8. $\dfrac{2}{8}$ $\dfrac{1}{4}$

Now re-read the title I CAN statement; decide whether you have 'Achieved' or need to revisit.

I CAN simplify fractions by finding a number that fits into both top and bottom

Simplifying a fraction means putting it into its lowest equivalent terms. To do this we find the BIGGEST number (factor) which fits into both top and bottom exactly. This is called the Highest Common Factor (HCF).

Look at $\frac{10}{15}$: what is the **biggest** number (factor) which will **fit into** both top and bottom?

To find numbers which fitted in, I tried out some Times Tables:
2, 4, 6, 8, ⑩ 12, 14, 16... ah, the 2x Table **won't** hit both!
3, 6, 9, 12, ⑮ 18... ah, the 3x Table **won't** hit both!
4, 8, 12, 16... ah, the 4x Table **won't** hit both!
5, ⑩ ⑮ 20... Yes! **5x Table will** hit both!

Then say, "How many times will **5** fit in to both top and bottom?"

It fits **2 times** into the 10   → $\frac{2}{3}$
... and **3 times** into the 15   →

TIP
I recommend using the 'pop-up-finger-method' and writing the steps of Times Tables out like this; it is good to count up and to see it visually.
With practice, you should find you don't always need to start with the 2x Table.

More examples:

$\frac{12}{20}$ → **4** is the biggest number to fit into both top and bottom. → $\frac{3}{5}$
(Also 2 would work, but 4 is bigger so go with 4.)
It fits 3x into 12... and 5x into 20.

$\frac{3}{9}$ → **3** is the biggest number to fit into both top and bottom. → $\frac{1}{3}$
3 fits 1x into 3... and 3x into 9.

$\frac{20}{50}$ → **10** is the biggest number to fit into both top and bottom. → $\frac{2}{5}$
(Also 2 and 5 would work, but 10 is biggest so go with 10.)
It fits 2x into 20... and 5x into 50.

Try

Simplify these fractions.
Find the BIGGEST number (factor) which will fit INTO the top and the bottom exactly:

$\frac{6}{9}$ $\frac{15}{20}$ $\frac{4}{10}$ $\frac{6}{12}$ $\frac{14}{20}$ $\frac{14}{21}$ $\frac{9}{12}$ $\frac{8}{12}$

Now make up 2 more similar questions of your own!

Test

Simplify these fractions.
Find the BIGGEST number (factor) which will fit INTO the top and the bottom exactly:

1. $\frac{6}{12}$

2, 3 and 6 will work; choose 6.

Simplifies to $\frac{1}{2}$

2. $\frac{14}{21}$

Only 7 will work.

Simplifies to $\frac{2}{3}$

3. $\frac{24}{40}$

2, 4 and 8 will work; choose 8.

Simplifies to $\frac{3}{5}$

4. $\frac{9}{12}$

Only 3 will work.

Simplifies to $\frac{3}{4}$

5. $\frac{2}{8}$

Only 2 will work.

Simplifies to $\frac{1}{4}$

6. $\frac{12}{15}$

Only 3 will work.

Simplifies to $\frac{4}{5}$

7. $\frac{6}{18}$

2, 3 and 6 will work; choose 6.

Simplifies to $\frac{1}{3}$

8. $\frac{10}{20}$

2, 5 and 10 will work; choose 10.

Simplifies to $\frac{1}{2}$

Now re-read the title I CAN statement; decide whether you have 'Achieved' or need to revisit.

I CAN simplify fractions by letting a Times Table jump out at me

 Knowing your Times Tables can help you simplify fractions. If a Times Table 'jumps out at you', you know which Times Table to count up in:

Look at this fraction: $\frac{21}{35}$

Which Times Table hits both **21** and **35**?

If you know your Times Tables well, this fraction is shouting: $\frac{21}{35}$

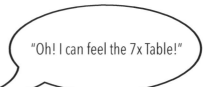
"Oh! I can feel the 7x Table!"

This is because **21** and **35** are both in the **7x Table**.

So all you need to do to simplify $\frac{21}{35}$ is to count up in sevens to see how many sevens fit into the top and bottom: 7, 14, ㉑, 28, ㉟, 42... How many sevens fit in to 21? **3**   How many sevens fit into 35? **5**

$\frac{21}{35}$ simplified is $\frac{3}{5}$

TIP
I recommend using the 'pop-up-finger-method' and writing the steps of Times Tables out, as is shown here; it is good to count up and to see it visually.

More examples:
If more than one Times Table shouts, use the higher one.
*Note how the **position** of the circled numbers relates to the simplified fraction.*

$\frac{18}{81}$:   9x Table jumps out   9, ⑱ 27, 36, 45, 54, 63, 72, ㉛..   Simplifies to $\frac{2}{9}$

$\frac{8}{20}$:   4x Table jumps out   4, ⑧ 12, 16, ⑳..   Simplifies to $\frac{2}{5}$

$\frac{36}{96}$:   12x Table jumps out   12, 24, ㊱ 48, 60, 72, 84, ㊉ 108, 120...   Simplifies to $\frac{3}{8}$

Try

Which Times Table jumps out at you? Go on to simplify your answers.

$\frac{8}{16}$ $\frac{16}{24}$ $\frac{27}{45}$ $\frac{35}{40}$ $\frac{14}{21}$ $\frac{15}{18}$ $\frac{6}{9}$ $\frac{2}{12}$

Now make up 2 more similar questions of your own!

Test

Which Times Table jumps out at you? Go on to simplify your answers.

1. $\frac{30}{35}$

    5x Table

    Simplifies to $\frac{6}{7}$

2. $\frac{14}{35}$

    7x Table

    Simplifies to $\frac{2}{5}$

3. $\frac{27}{81}$

    9x Table

    Reduces to $\frac{3}{9}$ Simplifies to $\frac{1}{3}$

4. $\frac{66}{77}$

    11x Table

    Simplifies to $\frac{6}{7}$

5. $\frac{15}{18}$

    3x Table

    Simplifies to $\frac{5}{6}$

6. $\frac{24}{144}$

    12x Table

    Reduces to $\frac{2}{12}$ Simplifies to $\frac{1}{6}$

7. $\frac{20}{24}$

    4x Table

    Simplifies to $\frac{5}{6}$

8. $\frac{16}{64}$

    8x Table

    Reduces to $\frac{2}{8}$ Simplifies to $\frac{1}{4}$

Now re-read the title I CAN statement; decide whether you have 'Achieved' or need to revisit.

I CAN choose an appropriate simplifying fractions method

We have looked at 3 ways to put a fraction into lower terms (simplify):
1. **chopping in half if both numbers are even**
2. **finding a number which fits into both top and bottom**
3. **letting a Times Table jump out.**

Try to have these **3 ways 'ready for action'** in your head, whenever working with fractions:

Say, "Is it CHOP-IN-HALF-ABLE?" **If yes, do it** as many times as you can. **If not**, find the **biggest number** (HCF) which will **fit into** the top and the bottom exactly, allowing a Times Table to jump out at you if possible.

Look at these fractions and see how they have been simplified:

$\frac{15}{25}$ → it is **not** CHOP-IN-HALF-ABLE, but the 5x Table jumps out, and 5 is the biggest number to fit into both top and bottom. Say, "How many times will 5 fit in to top and bottom?" → $\frac{3}{5}$

$\frac{6}{14}$ → it **is** CHOP-IN-HALF-ABLE → $\frac{3}{7}$

$\frac{15}{18}$ → it is **not** CHOP-IN-HALF-ABLE, but the 3x Table jumps out, and 3 is the biggest number to fit into both top and bottom. Say, "How many times will 3 fit in to top and bottom?" → $\frac{5}{6}$

$\frac{8}{20}$ → it **is** CHOP-IN-HALF-ABLE → $\frac{4}{10}$ ...and again → $\frac{2}{5}$

TIP
Sometimes a combination of the methods is needed, e.g.
$\frac{6}{18}$ → $\frac{3}{9}$ → $\frac{1}{3}$

Try

Simplify these fractions into their lowest terms:

$\frac{12}{20}$ $\frac{5}{15}$ $\frac{8}{16}$ $\frac{2}{8}$ $\frac{10}{15}$ $\frac{6}{10}$ $\frac{6}{9}$ $\frac{4}{8}$

Now make up 2 more similar questions of your own!

Test

Simplify these fractions into their lowest terms:

1. $\dfrac{8}{10}$ $\dfrac{4}{5}$

2. $\dfrac{4}{10}$ $\dfrac{2}{5}$

3. $\dfrac{6}{18}$ $\dfrac{1}{3}$

4. $\dfrac{12}{24}$ $\dfrac{1}{2}$

5. $\dfrac{10}{12}$ $\dfrac{5}{6}$

6. $\dfrac{12}{15}$ $\dfrac{4}{5}$

7. $\dfrac{15}{25}$ $\dfrac{3}{5}$

8. $\dfrac{5}{14}$ ... already in its lowest term! $\dfrac{5}{14}$

Now re-read the title I CAN statement; decide whether you have 'Achieved' or need to revisit.

I CAN add and subtract fractions with the same denominator

When adding or subtracting fractions, it is very straight forward as long as the denominators are the same:
- Add or subtract the top numbers (numerators) across.
- DO NOT add or subtract the bottom numbers (denominators) – they stay the same.

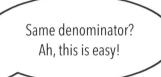

Same denominator? Ah, this is easy!

Look at these addition / subtraction of fractions examples to see how each answer is made:

$\frac{4}{7} + \frac{2}{7}$ → add the 4 and 2 to make the top 6, and the bottom stays the same → $\frac{6}{7}$

$\frac{3}{9} + \frac{2}{9}$ → add the 3 and 2 to make the top 5, and the bottom stays the same → $\frac{5}{9}$

$\frac{5}{8} - \frac{3}{8}$ → 5 minus 3 is 2 for the top, and the bottom stays the same → $\frac{2}{8}$ → $\frac{1}{4}$ (simplified)

TIP
Always simplify your answer.

Try

Add / subtract these fractions and simplify the answers where appropriate:

$\frac{2}{8} + \frac{1}{8}$ $\qquad$ $\frac{3}{4} + \frac{1}{4}$ $\qquad$ $\frac{7}{8} - \frac{2}{8}$ $\qquad$ $\frac{12}{20} - \frac{1}{20}$

$\frac{6}{7} - \frac{2}{7}$ $\qquad$ $\frac{3}{8} + \frac{3}{8}$ $\qquad$ $\frac{5}{6} - \frac{2}{6}$ $\qquad$ $\frac{8}{9} - \frac{3}{9}$

Now make up 2 more similar questions of your own!

Test

Add / subtract these fractions and simplify the answers where appropriate:

1. $\frac{2}{5} + \frac{1}{5}$ $\frac{3}{5}$

2. $\frac{3}{6} - \frac{1}{6}$ $\frac{2}{6}$ Simplifies to $\frac{1}{3}$

3. $\frac{1}{3} + \frac{1}{3}$ $\frac{2}{3}$

4. $\frac{5}{12} + \frac{5}{12}$ $\frac{10}{12}$ Simplifies to $\frac{5}{6}$

5. $\frac{5}{8} - \frac{1}{8}$ $\frac{4}{8}$ Simplifies to $\frac{1}{2}$

6. $\frac{4}{7} + \frac{3}{7}$ $\frac{7}{7}$ Simplifies to 1 whole

7. $\frac{7}{8} - \frac{2}{8}$ $\frac{5}{8}$

8. $\frac{10}{22} + \frac{1}{22}$ $\frac{11}{22}$ Simplifies to $\frac{1}{2}$

Now re-read the title I CAN statement; decide whether you have 'Achieved' or need to revisit.

I CAN add and subtract fractions with different denominators

When adding or subtracting fractions, it is very straight forward as long as the denominators are the same. But sometimes the denominators are NOT the same, so we need to make them the same.

How do we make denominators the same?
Find the **lowest common denominator (LCD)**,
as explained earlier in this Unit.

> Whatever you do to the top, you have to do to the bottom... whatever you do to the bottom, you have to do to the top!

Look at these addition / subtraction of fractions examples to see how each answer is made:

$\frac{2}{5} + \frac{3}{10}$ → Denominators are different, so find the LCD for 5 and 10 (which is 10) and re-write the sum so **both** fractions are out of 10

$\frac{2}{5} \, {}^{x2}_{x2}$ → $\frac{4}{10}$ So, $\frac{4}{10} + \frac{3}{10} = \frac{7}{10}$

$\frac{2}{3} - \frac{1}{4}$ → Denominators are different, so find the LCD for 3 and 4 (which is 12) and re-write the subtraction so both fractions are out of 12

$\frac{2}{3} \, {}^{x4}_{x4}$ → $\frac{8}{12}$
$\frac{1}{4} \, {}^{x3}_{x3}$ → $\frac{3}{12}$ So, $\frac{8}{12} - \frac{3}{12} = \frac{5}{12}$

$\frac{2}{5} + \frac{2}{6}$ → Denominators are different, so find the LCD for 5 and 6 (which is 30) and re-write the sum so both fractions are out of 30

$\frac{2}{5} \, {}^{x6}_{x6}$ → $\frac{12}{30}$
$\frac{2}{6} \, {}^{x5}_{x5}$ → $\frac{10}{30}$ So, $\frac{12}{30} + \frac{10}{30} = \frac{22}{30}$

(Simplifies to $\frac{11}{15}$)

Try

TIP
Always simplify your answer.

Add or subtract these fractions after making them have the same LCD.
Remember to simplify your answers:

$\frac{1}{3} + \frac{2}{4}$ $\frac{3}{10} + \frac{4}{5}$ $\frac{5}{6} - \frac{1}{10}$ $\frac{2}{3} - \frac{2}{5}$

$\frac{4}{5} + \frac{3}{8}$ $\frac{1}{2} - \frac{1}{7}$ $\frac{1}{2} + \frac{2}{3}$ $\frac{4}{5} - \frac{1}{4}$

Now make up 2 more similar questions of your own!

Test

Add or subtract these fractions after making them have the same LCD.
Remember to simplify your answers:

1. $\frac{1}{2} - \frac{1}{4}$

LCD = 4

$\frac{2}{4} - \frac{1}{4} = \frac{1}{4}$

2. $\frac{4}{5} - \frac{2}{8}$

LCD = 40

$\frac{32}{40} - \frac{10}{40} = \frac{22}{40}$

Simplifies to $\frac{11}{20}$

3. $\frac{1}{2} + \frac{1}{3}$

LCD = 6

$\frac{3}{6} + \frac{2}{6} = \frac{5}{6}$

4. $\frac{1}{5} + \frac{2}{4}$

LCD = 20

$\frac{4}{20} + \frac{10}{20} = \frac{14}{20}$

Simplifies to $\frac{7}{10}$

5. $\frac{5}{6} + \frac{2}{10}$

LCD = 30

$\frac{25}{30} + \frac{6}{30} = \frac{31}{30}$

Simplifies to $1\frac{1}{30}$

6. $\frac{9}{10} - \frac{4}{5}$

LCD = 10

$\frac{9}{10} - \frac{8}{10} = \frac{1}{10}$

7. $\frac{2}{3} + \frac{1}{4}$

LCD = 12

$\frac{8}{12} + \frac{3}{12} = \frac{11}{12}$

8. $\frac{4}{5} + \frac{1}{6}$

LCD = 30

$\frac{24}{30} + \frac{5}{30} = \frac{29}{30}$

102   Now re-read the title I CAN statement; decide whether you have 'Achieved' or need to revisit.

I CAN multiply fractions

Multiplying fractions is more straight forward than adding and subtracting them, as you don't have to worry about denominators being the same.

Look at $\frac{1}{3}$ x $\frac{4}{5}$ Simply **multiply the numerators** across, then **multiply the denominators** across: $\frac{1}{3}$ x $\frac{4}{5}$ = $\frac{4}{15}$

But also: 3 things to remember

1. If you see a **mixed number**, e.g. $3\frac{1}{5}$ **x** $\frac{2}{3}$, change it into an improper fraction **FIRST**, then re-write the question:

so becomes
$\frac{16}{5}$ x $\frac{2}{3}$

2. If you see a lone **whole number**, e.g. $\frac{2}{9}$ **x 4** , re-write it as a fraction by giving it a denominator of 1 (4 = $\frac{4}{1}$):

so becomes
$\frac{2}{9}$ x $\frac{4}{1}$

3. Look to see if it is possible to '**cross-cancel**' (simplify) **diagonals BEFORE** you multiply across:

e.g. $\frac{9}{10}$ x $\frac{5}{12}$

- Is there a number which **fits into** both ⑨ & ⑫? (3 fits into ⑨ three times & into ⑫ four times)

$_{2}^{3}\frac{9}{10}$ x $\frac{5}{12}_{4}^{1}$ = $\frac{3}{2}$ x $\frac{1}{4}$

- Is there a number which **fits into** both ⑩ & ⑤? (5 fits into ⑩ twice & into ⑤ once)

TIP
Always simplify your answer.

Try

Multiply these fractions. Remember to simplify your answers:

$\frac{3}{4}$ x $\frac{1}{2}$ $\frac{3}{16}$ x $\frac{4}{9}$ $1\frac{2}{3}$ x $1\frac{1}{3}$ 2 x $\frac{1}{3}$

$\frac{3}{8}$ x 2 $\frac{15}{17}$ x $\frac{1}{3}$ $\frac{3}{4}$ x $\frac{1}{10}$ $3\frac{1}{2}$ x $\frac{1}{3}$

Now make up 2 more similar questions of your own!

Test

Multiply these fractions. Remember to simplify your answers:

1. $\frac{3}{7}$ x $\frac{1}{2}$

$\frac{3}{14}$

2. $\frac{9}{15}$ x $\frac{5}{6}$

$\frac{3\cancel{9}}{3\cancel{15}} \times \frac{\cancel{5}^1}{\cancel{6}_2} = \frac{3}{6}$

Simplifies to $\frac{1}{2}$

3. $2\frac{1}{2}$ x $\frac{1}{3}$

$\frac{5}{2} \times \frac{1}{3} = \frac{5}{6}$

4. 4 x $\frac{2}{5}$

$\frac{4}{1} \times \frac{2}{5} = \frac{8}{5}$

Simplifies to $1\frac{3}{5}$

5. $\frac{4}{5}$ x $\frac{1}{3}$

$\frac{4}{15}$

6. $2\frac{1}{3}$ x $1\frac{1}{5}$

$\frac{7}{{}_1\cancel{3}} \times \frac{\cancel{6}^2}{5} = \frac{14}{5}$

Simplifies to $2\frac{4}{5}$

7. $\frac{4}{5}$ x 2

$\frac{4}{5} \times \frac{2}{1} = \frac{8}{5}$

Simplifies to $1\frac{3}{5}$

8. $\frac{4}{7}$ x $\frac{2}{3}$

$\frac{8}{21}$

Now re-read the title I CAN statement; decide whether you have 'Achieved' or need to revisit.

I CAN divide fractions

Dividing fractions is just as easy as multiplying them. However, there are two extra steps you need to know about – we need to 'flip' the second fraction (like flipping a pancake), and then change the symbol.

It is fun, quick and best of all, easier than it sounds! Look at $\frac{3}{7} \div \frac{1}{2}$

If you see a **division sign** in a fraction question, do these 2 things:

1. **'flip'** the **second** fraction (turn it on its head / turn it upside down)
2. change the **division** sign into a **multiplication** sign

Here is how it looks to start with: $\frac{3}{7} \div \frac{1}{2}$ which becomes $\frac{3}{7} \times \frac{2}{1}$ when re-written.

Then just follow **all** the multiplying fractions steps shown in the previous skill, treating it **exactly as you would a multiplication**:

$$\frac{3}{7} \times \frac{2}{1} = \frac{6}{7}$$

There is a fancy word for the **flip**: it is called the **reciprocal**. It is a great word! Say it out loud.

IMPORTANT
Make mixed numbers and whole numbers into improper fractions **before** you begin.

Try

Divide these fractions. Remember to simplify your answers:

$\frac{1}{3} \div \frac{1}{2}$ $\frac{2}{5} \div \frac{1}{4}$ $2\frac{1}{2} \div \frac{2}{3}$ $\frac{9}{12} \div \frac{3}{4}$

$\frac{3}{7} \div \frac{1}{2}$ $\frac{4}{7} \div \frac{1}{2}$ $\frac{9}{20} \div \frac{3}{5}$ $4 \div \frac{4}{7}$

Now make up 2 more similar questions of your own!

Test

Divide these fractions. Remember to simplify your answers:

1. $\dfrac{2}{7} \div \dfrac{1}{2}$ $\qquad\qquad$ $\dfrac{2}{7}$ x $\dfrac{2}{1}$ = $\dfrac{4}{7}$

2. $\dfrac{8}{15} \div \dfrac{2}{3}$ $\qquad\qquad$ $\dfrac{^4\cancel{8}}{_5\cancel{15}}$ x $\dfrac{\cancel{3}^1}{\cancel{2}_1}$ = $\dfrac{4}{5}$

3. $2\dfrac{2}{3} \div \dfrac{1}{2}$ $\qquad\qquad$ $\dfrac{8}{3}$ x $\dfrac{2}{1}$ = $\dfrac{16}{3}$

    Simplifies to $5\dfrac{1}{3}$

4. $\dfrac{4}{7} \div \dfrac{1}{4}$ $\qquad\qquad$ $\dfrac{4}{7}$ x $\dfrac{4}{1}$ = $\dfrac{16}{7}$

    Simplifies to $2\dfrac{2}{7}$

5. $3 \div \dfrac{2}{5}$ $\qquad\qquad$ $\dfrac{3}{1}$ x $\dfrac{5}{2}$ = $\dfrac{15}{2}$

    Simplifies to $7\dfrac{1}{2}$

6. $\dfrac{3}{8} \div \dfrac{1}{3}$ $\qquad\qquad$ $\dfrac{3}{8}$ x $\dfrac{3}{1}$ = $\dfrac{9}{8}$

    Simplifies to $1\dfrac{1}{8}$

7. $1\dfrac{2}{3} \div \dfrac{1}{4}$ $\qquad\qquad$ $\dfrac{5}{3}$ x $\dfrac{4}{1}$ = $\dfrac{20}{3}$

    Simplifies to $6\dfrac{2}{3}$

8. $\dfrac{2}{9} \div \dfrac{3}{7}$ $\qquad\qquad$ $\dfrac{2}{9}$ x $\dfrac{7}{3}$ = $\dfrac{14}{27}$

Now re-read the title I CAN statement; decide whether you have 'Achieved' or need to revisit.

I CAN make an alarm bell go off in my brain for 'half' when I see 0.5 or 50%

Make a loud ALARM BELL ring in your head when you see:

0.5   0.50   0.500   50%

Whole numbers 50 and 500 get involved too, e.g. for length, capacity and weight.

It sounds complicated when we read maths instructions like these:

Find **50%** of...   What is **0.5** of...?   What is **50%** of...?

But if your alarm bell rings loudly, you know it simply means **half** ($\frac{1}{2}$).

See how the 'half' alarm bell can help your thinking:

50% of 70 dogs.....................................half of 70 dogs is **35 dogs**
0.5 of a litre..half of 1 L (1,000 mL) is **500 mL**
50% of £3...half of £3 is **£1.50**
500 g is what fraction of a kilogram?.....**half** ($\frac{1}{2}$) because half a kg (1,000 g) is 500 g

Try

Work out the answers to these fractions, decimals and percentages questions by making the 'just means half!' alarm bell go off in your head:

50% of 120 cats wear collars. How many cats wear collars?

Draw a circle and shade 50%

There are 1,000 millilitres (mL) in a litre (L). How many millilitres in 0.5 of a litre?

A calculator reads: 0.5   What does this mean?

In a school there were 60 children. If 50% were girls, how many were boys?

Find 0.5 of 42 eggs

50% of £20 is...?

A calculator reads: 0.5000000   What does this mean?

Now make up 2 more similar questions of your own!

Test

Work out the answers to these fractions, decimals and percentages questions by making the 'just means half!' alarm bell go off in your head:

1.	What is 50% of £8?	£4
2.	0.5 of 6 boys is...?	3 boys
3.	50% of 60 pencils need sharpening – how many is this?	30 pencils
4.	50% of 400 cats are male. How many are male?	200 cats
5.	Find 0.50 of 12 fish	6 fish
6.	50% of the 12 trees in my garden are evergreens. How many is this?	6 trees
7.	What is 0.50 of 80?	40
8.	There is 50% off all T shirts in the store – today only! T shirts are usually £30. What is the sale price?	£15

Now re-read the title I CAN statement; decide whether you have 'Achieved' or need to revisit.

I CAN make an alarm bell go off in my brain for 'quarter' when I see 0.25 or 25%

Make a loud ALARM BELL ring in your head when you see:

0.25   **0.250**   **25%**

Whole numbers 25 and 250 get involved too, e.g. for length, capacity and weight.

It sounds complicated when we read maths instructions like these:

Find **25%** of...   What is **0.25** of...?   What is **25%** of...?

But if your alarm bell rings loudly, you know it simply means **a quarter** ($\frac{1}{4}$).
Remember: to find a quarter, you just half it... then half it again.

See how the 'quarter' alarm bell can help your thinking:

25% of 80 apples.................................a quarter of 80 apples is **20 apples**
0.25 of a litre ....a quarter of 1 L (1,000 mL) is **250 mL**
25% of £10...a quarter of £10 is **£2.50**
250 m is what fraction of a km?............one **quarter** ($\frac{1}{4}$) because a quarter
                                                                         of a km (1,000 m) is 250 m

Try

Work out the answers to these fractions, decimals and percentages questions by making the 'just means a quarter!' alarm bell go off in your head:

0.25 of 12 fish?

25% of 200 pencils were broken – how many is this?

25% of 80 dogs?

Draw a rectangle and shade 25% red

0.25 is a decimal – express as a fraction

A litre is 1,000 mL – how many mL in a quarter of a litre?

What is 25% of 60?

Find 0.25 of 4 elephants

Now make up 2 more similar questions of your own!   **109**

Test

Work out the answers to these fractions, decimals and percentages questions by making the 'just means a quarter!' alarm bell go off in your head:

1.	Find 0.25 of 40	10
2.	What is 25% of 800?	200
3.	Express 0.25 as a fraction in its lowest form	$\frac{1}{4}$
4.	What is 0.25 of £100?	£25
5.	There are 20 lions. 25% of them are sleeping. How many are sleeping?	5 lions
6.	There are 12 hats on a stand and 25% of them are black. How many is this?	3 hats
7.	What is 0.25 of 16?	4
8.	25% of 8 dogs are brown. How many is this?	2 dogs

Now re-read the title I CAN statement; decide whether you have 'Achieved' or need to revisit.

I CAN make an alarm bell go off in my brain for 'three-quarters' when I see 0.75 or 75%

Make a loud ALARM BELL ring in your head when you see:

0.75   **0.750**   **75%**

Whole numbers 75 and 750 get involved too, e.g. for length, capacity and weight.

It sounds complicated when we read maths instructions like these:

Find **75%** of...   What is **0.75** of...?   What is **75%** of...?

But if your alarm bell rings loudly, you know it simply means **three-quarters** ($\frac{3}{4}$).
How do we find three-quarters? DO NOT try to find three-quarters, instead, find one quarter first. Once you have **one** quarter, you need **three** of them, so multiply by 3.

See how the 'three-quarters' alarm bell can help your thinking:

75% of £20.................half it then half it again to **find one quarter (£5)**,
                            then multiply that by 3: **three-quarters** of £20 is **£15**
0.75 of 400 ....half it then half it again to **find one quarter (100)**,
                            then multiply that by 3: **three-quarters** of 400 is **300**
$\frac{3}{4}$ **of 40 cats**..............half it then half it again to **find one quarter (10)**,
                            then multiply that by 3: **three-quarters** of 40 cats is **30 cats**
75% of 800 fish ....half it then half it again to **find one quarter (200)**,
                            then multiply that by 3: **three-quarters** of 800 fish is **600 fish**

Try

Work out the answers to these fractions, decimals and percentages questions by making the 'just means three-quarters!' alarm bell go off in your head:

Find 75% of £8   What is 0.75 of 12 eggs?
Find 0.75 of 40 girls   Express $\frac{3}{4}$ as a decimal and a percentage
Find 75% of one pound   75% of 400 is...?
What is 0.75 of a litre?   75% of 60 shops were open. How many is this?

Now make up 2 more similar questions of your own!

Test

Work out the answers to these fractions, decimals and percentages questions by making the 'just means three-quarters!' alarm bell go off in your head:

1.	75% of 100 flowers were red. How many is this?	75 flowers

2. Draw a circle, divide it into quarters, then shade 75%   e.g.

3.	What is 0.75 of 60?	45

4. Express three-quarters as a decimal and a percentage   0.75 and 75%

5.	What is 75% of 400 pencils?	300 pencils

6. What is 0.75 of a metre?   75 cm (or 750 mm)

7.	What is 0.75 of a km?	750 m

8. Find 75% of 80   60

Now re-read the title I CAN statement; decide whether you have 'Achieved' or need to revisit.

I KNOW my 25x Table to 100

Oh good – this is a quickie: can you count in 25s?
Count up in 25s until you hit 100. Say over and over:

25   50   75   100

Knowing the first 4 steps of the **25x Table** is helpful for 2 reasons:

1. It pops up all the time in **fractions, decimals and percentages** work
2. It pops up all the time in **measurement** work

Look at these measurements for **capacity, weight** and **distance**.
Say each one out loud and see if you can spot a **25x Table** connection or pattern.

Notice the extra zeros:

L and mL	kg and g	km and m
1 L = **100**0 mL	1 kg = **100**0 g	1 km = **100**0 m
$\frac{3}{4}$ L = **750** mL	$\frac{3}{4}$ kg = **750** g	$\frac{3}{4}$ km = **750** m
$\frac{1}{2}$ L = **500** mL	$\frac{1}{2}$ kg = **500** g	$\frac{1}{2}$ km = **500** m
$\frac{1}{4}$ L = **250** mL	$\frac{1}{4}$ kg = **250** g	$\frac{1}{4}$ km = **250** m

Try

Use the first four 25x Table facts to help you work through these:

Say your 25x Table from zero to 100 in a loud voice!

Say your 25x Table from 100 to zero (so backwards) in a loud voice!

How many mL are there in $\frac{1}{2}$ L?

How many g are there in $\frac{3}{4}$ kg?

What is missing in the number pattern: 25, 50, ___ , 100?

How many m in $\frac{1}{4}$ km?

Count up in 250s until you reach 1,000

How many g are there in $\frac{1}{2}$ kg?

Now make up 2 more similar questions of your own!

Test

Use the first four 25x Table facts to help you work through these:

1.	Say the 25x Table up to 100 out loud	25, 50, 75, 100
2.	Starting at 100, count back in 25s until you reach zero	100, 75, 50, 25, 0
3.	How many mL are there in half a litre?	500 mL
4.	How many metres are there in $\frac{3}{4}$ km?	750 m
5.	What is $\frac{1}{2}$ (50% or 0.5) of a litre?	500 mL
6.	Count up in 250s until you reach 1000	250, 500, 750, 1000
7.	How many 25s are there in 75?	3
8.	How many grams in $\frac{3}{4}$ kg?	750 g

Now re-read the title I KNOW statement; decide whether you have 'Achieved' or need to revisit.

I CAN apply fractions, decimals and percentages alarm bells knowledge to real life maths problems

 When we get into harder fractions, decimals and percentages work, we often forget there are some tricks and alarm bells for spotting $\frac{1}{4}$, $\frac{1}{2}$ and $\frac{3}{4}$.

Spot the three important alarm bells:

25%	0.25	250	Did you hear a **quarter** alarm bell?
50%	0.50	500	Did you hear a **half** alarm bell?
75%	0.75	750	Did you hear a **three-quarters** alarm bell?

length / distance:
A metre has 100 centimetres

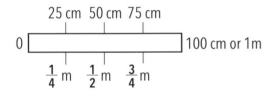

length / distance:
A kilometre has 1,000 metres

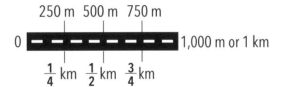

capacity / liquid:
A litre has 1,000 mL

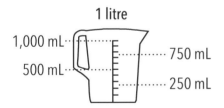

weight / mass:
A kilogram has 1,000 g

$250\,g = \frac{1}{4}$ kg

$500\,g = \frac{1}{2}$ kg

$750\,g = \frac{3}{4}$ kg

$1{,}000\,g = 1$ kg

Try

See if you can apply the fractions, decimals and percentages alarm bells to fill the blanks:

0.5 of 1 m is _____ cm

50% of 1 L is _____ mL

0.75 of 60 pencils is _____ pencils

25% of 1 m is _____ cm

75% of 1 km is _____ m

0.25 of 1 km is _____ m

$\frac{1}{4}$ of 1 kg is _____ g

0.5 of 80 dogs is _____ dogs

Now make up 2 more similar questions of your own!

Test

See if you can apply the fractions, decimals and percentages alarm bells to fill the blanks:

1. 0.25 of 1 m is _____ cm   25 cm

2. 50% of 100 cakes is _____ cakes   50 cakes

3. 75% of 1 kg is _____ g   750 g

4. 50% of 1 L is _____ mL   500 mL

5. 25% of 800 dogs is _____ dogs   200 dogs

6. 0.5 of a kilometre is _____ m   500 m

7. 0.75 of a litre is _____ mL   750 mL

8. 0.25 of 1 kilogram is _____ g   250 g

Now re-read the title I CAN statement; decide whether you have 'Achieved' or need to revisit.

I CAN find simple fractions of numbers mentally (where the numerator is one)

 We learned how to find a fraction of a number using the paper and pencil method for multiplying fractions. Here is another useful idea to try out.

Sometimes we need to find a **fraction of a number mentally**.
It is easy if the numerator is **one**:

e.g. Find $\frac{1}{3}$ of **21**   Find $\frac{1}{5}$ of **25**   Find $\frac{1}{6}$ of **18**

Count up (using the 'pop-up-finger-method') in whatever the **denominator** is, until you reach the number.

$\frac{1}{3}$ **of 21**: count up in **3s** until you hit **21**:   3, 6, 9, 12, 15, 18, ㉑, 24: **7**

$\frac{1}{5}$ **of 25**: count up in **5s** until you hit **25**:   5, 10, 15, 20, ㉕, 30: **5**

$\frac{1}{6}$ **of 18**: count up in **6s** until you hit **18**:   6, 12, ⑱ 24: **3**

This will work for **any** Times Table, but only consider using this method if the amount is less than 144 (12 x 12); if the number is greater than 144, e.g. $\frac{1}{3}$ of 354, counting up in threes will take too long!

TIP
If the fraction is
$\frac{1}{4}$ $\frac{1}{2}$ or $\frac{3}{4}$
remember you know a faster 'alarm bell' method.

Try

Count up in the relevant Times Table to find fractions of numbers:

$\frac{1}{7}$ of 21 $\frac{1}{3}$ of 18 $\frac{1}{5}$ of 30 $\frac{1}{3}$ of 9

$\frac{1}{3}$ of 27 $\frac{1}{7}$ of 14 $\frac{1}{6}$ of 48 $\frac{1}{8}$ of 16

Now make up 2 more similar questions of your own!

Test

Count up in the relevant Times Table to find fractions of numbers:

1. $\frac{1}{3}$ of 18   6

2. $\frac{1}{5}$ of 45   9

3. $\frac{1}{6}$ of 12   2

4. $\frac{1}{7}$ of 21   3

5. $\frac{1}{10}$ of 50   5

6. $\frac{1}{5}$ of 10   2

7. $\frac{1}{3}$ of 21   7

8. $\frac{1}{5}$ of 15   3

Now re-read the title I CAN statement; decide whether you have 'Achieved' or need to revisit.

I CAN find harder fractions of numbers mentally (where the numerator is more than one)

We learned how to find a fraction of a number using the paper and pencil method for multiplying fractions. Here is another useful idea to try out.

Sometimes we need to find a **fraction of a number mentally** where the numerator is **more** than one:

e.g. Find $\frac{2}{3}$ of...   Find $\frac{3}{5}$ of...   Find $\frac{5}{6}$ of...

An easy way to work these out, is to find **one portion** (**one** third, fifth or sixth) first:

> Don't find **two** thirds, find one **third** first!

Find $\frac{2}{3}$ **of 21** → $\frac{1}{3}$ of 21 is 7

Once you have $\frac{1}{3}$, multiply it by the numerator (x2): 7 x 2 = **14**

$\frac{2}{3}$ of 21 = 14

> Don't find **three** fifths, find **one** fifth first!

Find $\frac{3}{5}$ **of 25** → $\frac{1}{5}$ of 25 is 5

Once you have $\frac{1}{5}$, multiply it by the numerator (x3): 5 x 3 = **15**

$\frac{3}{5}$ of 25 = 15

> Don't find **five** sixths, find **one** sixth first!

Find $\frac{5}{6}$ **of 24** → $\frac{1}{6}$ of 24 is 4

Once you have $\frac{1}{6}$, multiply it by the numerator (x5): 4 x 5 = **20**

$\frac{5}{6}$ of 24 = 20

Try

Solve these fractions questions, by finding ONE portion first, then multiplying by the numerator:

$\frac{2}{3}$ of 18 $\frac{2}{5}$ of 15 $\frac{5}{6}$ of 30 $\frac{4}{5}$ of 20

$\frac{2}{3}$ of 9 $\frac{3}{5}$ of 40 $\frac{2}{6}$ of 12 $\frac{3}{10}$ of 20

Now make up 2 more similar questions of your own!

Test

Solve these fraction questions, by finding ONE portion first, then multiplying by the numerator:

1. $\frac{2}{3}$ of 18

$\frac{1}{3} = 6$

$\frac{2}{3} = 12$

2. $\frac{5}{6}$ of 12

$\frac{1}{6} = 2$

$\frac{5}{6} = 10$

3. $\frac{2}{5}$ of 25

$\frac{1}{5} = 5$

$\frac{2}{5} = 10$

4. $\frac{3}{5}$ of 15

$\frac{1}{5} = 3$

$\frac{3}{5} = 9$

5. $\frac{2}{3}$ of 24

$\frac{1}{3} = 8$

$\frac{2}{3} = 16$

6. $\frac{3}{7}$ of 21

$\frac{1}{7} = 3$

$\frac{3}{7} = 9$

7. $\frac{5}{7}$ of 21

$\frac{1}{7} = 3$

$\frac{5}{7} = 15$

8. $\frac{4}{5}$ of 35

$\frac{1}{5} = 7$

$\frac{4}{5} = 28$

Now re-read the title I CAN statement; decide whether you have 'Achieved' or need to revisit.

I KNOW where tenths and hundredths live in the decimal place value grid

Tenths and hundredths: you must know place value to 2 decimal places in particular. Look at the place value grid and read the headings.

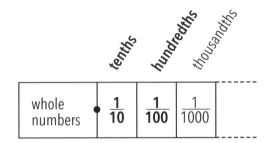

whole numbers	$\frac{1}{10}$	$\frac{1}{100}$	$\frac{1}{1000}$

Copy this grid onto paper. Be sure you know where the **tenths** and **hundredths** live. We rarely refer to the **thousandths** column, but it is important to know where it is, e.g. when working with kg/g, L/mL and km/m.

Remember that the '**ths**' at the end of the words **ten**, **hundred** and **thousand** makes them into fractions / parts of whole ones.

Look at how these numbers are broken down into wholes and decimals. Say them out loud:

45.7 → 45 whole ones and 7 ten**ths** → $45\frac{7}{10}$

8.04 → 8 whole ones and 4 hundred**ths** → $8\frac{4}{100}$

38.65 → 38 whole ones, 6 ten**ths** and 5 hundred**ths** → $38\frac{65}{100}$
or 38 wholes and 65 hundred**ths**

TIP
You could use this opportunity to incorporate the whole numbers place value grid (in Unit 1) as well, so your grid shows both whole and decimal place values.

Try

Check you are 100% sure which decimal place column is tenths, and which is hundredths:

| | | ? | Which column (whole numbers / tenths / hundredths) is this? |

| 4 | 8 | 2 | What is the 2 worth? |

| ? | | | Which column (whole numbers / tenths / hundredths) is this? |

| | | | Write $2\frac{3}{100}$ (2 whole ones and 3 hundredths) in the grid |

| 13 | 7 | 2 | What is the 7 worth? |

| | ? | | Which column (whole numbers / tenths / hundredths) is this? |

| 2 | 6 | 3 | What is the 6 worth? |

| | | | Write $\frac{42}{100}$ (42 hundredths) in the grid |

Now make up 2 more similar questions of your own!

Test

Work through these tenths and hundredths questions:

1. What is the 7 worth in 5.07?

7 hundredths or $\frac{7}{100}$

2. Write in the grid 12 whole ones and 9 tenths

	.	

12 . 9

3. 4.78... what is the 7 worth?

7 tenths or $\frac{7}{10}$

4. 16.38... what is the 8 worth?

8 hundredths or $\frac{8}{100}$

5. Write in the grid 2 whole ones and 3 hundredths

	.	

2 . 0 3

6. Write in the grid 1 whole one and 19 hundredths

	.	

1 . 1 9

7. What is the 2 worth?

4 . 2 6

2 tenths or $\frac{2}{10}$

8. What is the 5 worth?

8 . 6 5

5 hundredths or $\frac{5}{100}$

Now re-read the title I KNOW statement; decide whether you have 'Achieved' or need to revisit.

I KNOW that tenths are bigger than hundredths and can visualise them in a hundred square

This concept confuses many people!
Which is bigger – a tenth or a hundredth? What do you think?

Picture a pizza cut up into **tenths**: ten slices / parts / pieces.
Picture a pizza cut up into **hundredths**: a hundred slices / parts / pieces.

Does that help your thinking? Which fraction of pizza would you choose – a tenth or a hundredth?

Now look at these equally-sized grids. One shows a **tenth** shaded and the other shows a **hundredth** shaded:

tenth

$\frac{1}{10}$

hundredth

$\frac{1}{100}$

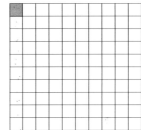

Now visualise these three yummy examples:

- Picture a slab of chocolate. Would you like a tenth or a hundredth? A **tenth** of a bar of chocolate is bigger than a **hundredth** of a bar of chocolate!

- $\frac{5}{10}$ of a pizza (this would be half) is bigger that $\frac{5}{100}$ of a pizza.

- $\frac{9}{10}$ of a bar of chocolate (that is almost all of it) is bigger than $\frac{9}{100}$ of a bar of chocolate.

Try

Using your knowledge of tenths and hundredths place value, decide which is the BIGGER of these decimal fractions:

0.04	or	0.4	1.7	or	1.07
3.02	or	3.2	10.5	or	10.05
2.30	or	2.03	1.05	or	1.50
1.09	or	1.90	2.40	or	2.04

Now make up 2 more similar questions of your own!

Test

Using your knowledge of tenths and hundredths place value, decide which is the BIGGER of these decimal fractions:

1.	0.01 or 0.1	0.1
2.	8.04 or 8.4	8.4
3.	7.02 or 7.20	7.20
4.	4.50 or 4.5	they are same
5.	2.7 or 2.07	2.7
6.	3.80 or 3.08	3.80
7.	4.600 or 4.6	they are same
8.	3.01 or 3.1	3.1

Now re-read the title I KNOW statement; decide whether you have 'Achieved' or need to revisit.

I KNOW that one tenth is the same as ten hundredths

Look at these 2 fractions and say them out loud: $\frac{1}{10}$ and $\frac{10}{100}$
They are equivalent fractions. Notice the extra zeros on the second fraction.

Look at these equally-sized grids. One shows a **tenth** strip shaded and the other shows **ten hundredths** shaded:

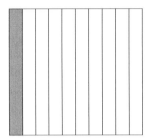   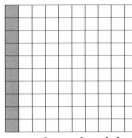

One tenth   is the same as   **ten hundredths**.   They are **equivalent**.

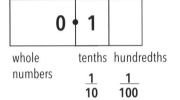   

Say the following statements out loud. Think about what they mean and refer to the above grids to help visualise and understand:

0.1 means one tenth ($\frac{1}{10}$); this means one out of ten parts.
0.10 means 10 hundredths ($\frac{10}{100}$); this means ten out of one hundred parts.
Therefore **0.1** and **0.10** represent the same value; they are **equivalent**.
0.100000000 etc. would be the same because the extra zeros do not affect the value.

Try

Refer to the grids above. Find the equivalent fractions:

$$\frac{2}{10} = \frac{?}{100} \qquad \frac{50}{100} = \frac{?}{10} \qquad \frac{3}{10} = \frac{?}{100} \qquad \frac{10}{100} = \frac{?}{10} \qquad \frac{7}{10} = \frac{?}{100} \qquad \frac{80}{100} = \frac{?}{10}$$

Four tenths is equivalent to how many hundredths?
Sixty hundredths is equivalent to how many tenths?

Now make up 2 more similar questions of your own!

Test

Find the equivalent fractions:

1. $\frac{5}{10} = \frac{?}{100}$ $\frac{50}{100}$

2. $\frac{60}{100} = \frac{?}{10}$ $\frac{6}{10}$

3. $\frac{3}{10} = \frac{?}{100}$ $\frac{30}{100}$

4. Forty hundredths is equivalent to how many tenths? $\frac{4}{10}$

5. $\frac{20}{100} = \frac{?}{10}$ $\frac{2}{10}$

6. $\frac{80}{100} = \frac{?}{10}$ $\frac{8}{10}$

7. $\frac{1}{10} = \frac{?}{100}$ $\frac{10}{100}$

8. Seven tenths is equivalent to how many hundredths? $\frac{70}{100}$

Now re-read the title I KNOW statement; decide whether you have 'Achieved' or need to revisit.

I CAN order or compare decimals by making them into fractions out of 100

We often need to order or compare two or more decimals. It is about working out which decimal is biggest or smallest.

It might say:

Which decimal is smallest / biggest...

Order the decimals from greatest to least or least to greatest...

Put the decimals in order starting with the smallest / biggest...

Order the decimals using < or >...

A method which ALWAYS works, is to make them **all** into fractions which have the **same denominator of 100**. Then you can easily see which is biggest / smallest.

Which is the biggest decimal here: **0.02, 0.22 or 0.2**? It is tempting to guess. **Do not guess**. Instead, **make them into fractions out of 100**.

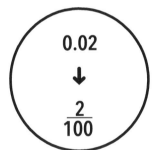

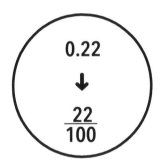

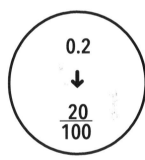

Now you can **easily** see which is the biggest decimal.

Try

Find which decimal is biggest in each set by making them into fractions out of 100:

0.4	0.44	0.04	2.01	2.10	2.11
4.05	4.5	4.55	3.6	3.66	3.06
3.4	3.44	3.04	2.19	2.91	2.9
2.8	2.98	2.88	1.7	1.77	1.07

Now make up 2 more similar questions of your own!

Test

Find which decimal is biggest in each set by making them into fractions out of 100:

1.	9.02	9.22	9.20		9.22
2.	3.4	3.44	3.04		3.44
3.	5.30	5.03	5.33		5.33
4.	6.1	6.01	1.66		6.1
5.	2.2	0.2	2.0		2.2
6.	1.7	1.77	1.07		1.77
7.	10.6	10.06	1.66		10.6
8.	3.08	8.80	3.88		8.80

Now re-read the title I CAN statement; decide whether you have 'Achieved' or need to revisit.

I CAN round decimals to the nearest tenth (one decimal place or 1 dp)

For decimals, we are often asked to 'round to the nearest tenth' or 'round to one decimal place' (1 dp).

This means that the decimal section needs to show **only** the **tenths** column.

Look at this place value grid and check you know the places confidently.

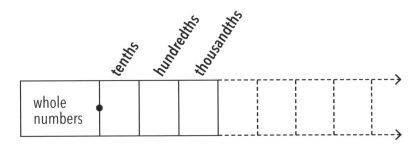

How to round to the **nearest tenth**:
1. Point to the number in the tenths column and hold your finger there.
2. Look at the **column to its right (the hundredths)**: if the digit there is 5 or more, round the tenths column up one; if it is less than 5, the digit in the tenths column stays the same.
3. Drop / lose all the digits after the tenths column.

Apply the steps above to work through these examples:

Round **4.72** to the nearest tenth (1 dp): **4.7** because the 2 (hundredths) is less than 5
Round **5.36** to the nearest tenth (1 dp): **5.4** because the 6 (hundredths) is more than 5
Round **49.6582** to the nearest tenth (1 dp): **49.7** because the 5 (hundredths) is exactly 5

Try

Round these decimals to the nearest tenth (one decimal place):

0.28	0.534	2.87	14.815
6.55	12.4871	1.06	12.54

Now make up 2 more similar questions of your own!

Test

Round these decimals to the nearest tenth (one decimal place):

1.	7.58	7.6
2.	0.43	0.4
3.	12.8765	12.9
4.	3.52	3.5
5.	19.758	19.8
6.	2.06	2.1
7.	3.21	3.2
8.	4.55	4.6

Now re-read the title I CAN statement; decide whether you have 'Achieved' or need to revisit.

I CAN round decimals to the nearest hundredth (two decimal places or 2 dp)

For decimals, we are often asked to 'round to the nearest hundredth' or 'round to two decimal places' (2 dp).

This means that the decimal section needs to show the **tenths and hundredths** columns.

Look at this place value grid and check you know the places confidently.

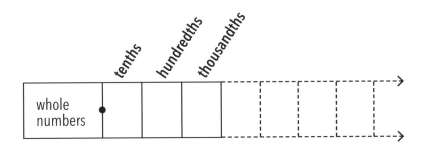

How to round to the **nearest hundredth**:
1. Point to the number in the hundredths column and hold your finger there.
2. Look at the **column to its right (the thousandths)**: if the digit there is 5 or more, round the hundredths column up one; if it is less than 5, the number in the hundredths column stays the same.
3. Drop / lose all the digits after the hundredths column.

Apply the steps above to work through these examples:

Round **5.374** to the nearest hundredth (2 dp): **5.37** because the 4 (thousandths) is less than 5
Round **2.826** to the nearest hundredth (2 dp): **2.83** because the 6 (thousandths) is more than 5
Round **26.37541** to the nearest hundredth (2 dp): **26.38** because the 5 (thousandths) is exactly 5

Try

Round these decimals to the nearest hundredth (two decimal places):

12.463	1.345	2.789	16.4445
1.485	0.233	14.726457	2.333

Now make up 2 more similar questions of your own!

Test

Round these decimals to the nearest hundredth (two decimal places):

1.	3.445	3.45
2.	12.328	12.33
3.	0.64482	0.64
4.	12.123	12.12
5.	4.685	4.69
6.	6.4467	6.45
7.	13.139	13.14
8.	1.722	1.72

Now re-read the title I CAN statement; decide whether you have 'Achieved' or need to revisit.

I CAN multiply decimals by 10, 100 or 1000 by making the decimal point jump to the right

 Think about what multiplying (by more than 1) does to a number. Does it make it bigger or smaller? It makes it bigger of course.

When we **multiply decimals** by a number greater than 1, they get **bigger**.

Here is a clever trick to multiply decimals by 10, 100 or 1000 using kangaroo jumps.

Think of the kangaroo being the **decimal point**:

- to multiply by 1**0**: the kangaroo jumps **one place to the right**
- to multiply by 1**00**: the kangaroo jumps **two places to the right**
- to multiply by 1**000**: the kangaroo jumps **three places to the right**

See the pattern? The number of zeros equals the number of jumps!

Follow the number of jumps with your finger:

463.24 x 10   **8.368 x 100**   **58.3748 x 1000**

I jump one place to the right

I jump two places to the right

I jump three places to the right

4632.4   **836.8**   **58374.8**

Try

Multiply these decimals:

53.267 x 10	4.7854 x 1000	55.216 x 100	104.27 x 10
12.750 x 100	5.3672 x 1000	0.4685 x 10	13.280 x 100

Now make up 2 more similar questions of your own!

Test

Multiply these decimals:

1.   5.7862 x 100   578.62

2.   202.46 x 10   2024.6

3.   13.4685 x 1000   13468.5

4.   2.50 x 10   25.0 (or just 25)

5.   8.703 x 100   870.3

6.   1.7541 x 1000   1754.1

7.   916.567 x 100   91656.7

8.   12.52 x 10   125.2

Now re-read the title I CAN statement; decide whether you have 'Achieved' or need to revisit.

I CAN divide decimals by 10, 100 or 1000 by making the decimal point jump to the left

 Think about what dividing (by more than 1) does to a number. Does it make it bigger or smaller? It makes it smaller of course.

When we **divide decimals** by a number greater than 1, they get **smaller**.

Here is a clever trick to divide decimals by 10, 100 or 1000 using kangaroo jumps.

Think of the kangaroo being the **decimal point**:

- to divide by 1**0**: the kangaroo jumps **one place to the left**
- to divide by 1**00**: the kangaroo jumps **two places to the left**
- to divide by 1**000**: the kangaroo jumps **three places to the left**

See the pattern? The number of zeros equals the number of jumps!

Follow the number of jumps with your finger:

463.24 ÷ 10   57.36 ÷ 100   5463.2 ÷ 1000

I jump one place to the left

I jump two places to the left

I jump three places to the left

46.324   0.5736   5.4632

Try

Divide these decimals:

53.269 ÷ 10   36724.78 ÷ 1000   815.21 ÷ 100   34.2 ÷ 10

536.01 ÷ 100   3612.75 ÷ 1000   84.687 ÷ 100   713.28 ÷ 10

Now make up 2 more similar questions of your own!

Test

Divide these decimals:

1. $578.62 \div 100$ 5.7862

2. $2024.6 \div 10$ 202.46

3. $13468.5 \div 1000$ 13.4685

4. $25.0 \div 10$ 2.50
(or just 2.5)

5. $870.3 \div 100$ 8.703

6. $1754.1 \div 1000$ 1.7541

7. $91656.7 \div 100$ 916.567

8. $125.2 \div 10$ 12.52

Now re-read the title I CAN statement; decide whether you have 'Achieved' or need to revisit.

I CAN change a fraction into a decimal or percentage by making it out of 100 (if the denominator fits exactly into 100)

We are often asked to express a fraction as a decimal or a percentage. This is easy if the fraction can be made out of 100 ($\frac{?}{100}$).

$\frac{4}{20}$ — *We can have ANY denominator! We are not fussy!*

Fractions are very easy going

0.2 — *We like being out of 10 or 100. Fuss, fuss...*

Decimals are fussy

20% — *We insist that we are out of 100! Fuss, fuss...*

Percentages are really fussy

So if we aim to have them **all** out of **100**, then everyone is happy. A fraction can be made to be **out of 100 easily** if the **denominator fits exactly into 100 without remainders.**

These **easy to handle denominators** all fit into 100 exactly: **2, 4, 5, 10, 20, 25, 50.** Write them out so you remember them, as they are very useful. **Say** how many times each will fit into 100.

- If a fraction is **out of 100 already**, it is ready to go, e.g. $\frac{45}{100} = 0.45 = 45\%$

- If a fraction is **not out of 100**, but has one of the 'easy to handle' denominators above, then make it an equivalent fraction with denominator 100,

e.g. $\frac{3}{50}$ Remember the rap: → $\frac{3}{50}{}^{\times 2}_{\times 2} = \frac{6}{100} = 0.06 = 6\%$

Whatever you do to the top, you have to do to the bottom... whatever you do to the bottom, you have to do to the top!

If a denominator will **not** fit exactly into a hundred without remainders, this method will not work, but we will learn how to use a calculator method next.

Try

Change these fractions into decimals and percentages by making them out of 100:

$\frac{9}{20}$ $\frac{23}{50}$ $\frac{3}{10}$ $\frac{2}{5}$ $\frac{6}{25}$ $\frac{4}{50}$ $\frac{2}{20}$ $\frac{7}{10}$

Now make up 2 more similar questions of your own!

Test

Change these fractions into decimals and percentages by making them out of 100:

1. $\frac{6}{20}$ (x 5) $\frac{30}{100}$ 0.30 / 30%

2. $\frac{4}{25}$ (x 4) $\frac{16}{100}$ 0.16 / 16%

3. $\frac{7}{10}$ (x 10) $\frac{70}{100}$ 0.70 / 70%

4. $\frac{2}{5}$ (x 20) $\frac{40}{100}$ 0.40 / 40%

5. $\frac{43}{50}$ (x 2) $\frac{86}{100}$ 0.86 / 86%

6. $\frac{1}{20}$ (x 5) $\frac{5}{100}$ 0.05 / 5%

7. $\frac{3}{25}$ (x 4) $\frac{12}{100}$ 0.12 / 12%

8. $\frac{27}{50}$ (x 2) $\frac{54}{100}$ 0.54 / 54%

Now re-read the title I CAN statement; decide whether you have 'Achieved' or need to revisit.

I CAN use a calculator to change a fraction into a decimal or percentage

We often need to express a fraction as a decimal or percentage. If the denominator does NOT fit into 100 equally to make it easy for us, we can use a calculator method.

We simply divide the **top by the bottom**.

e.g. $\frac{5}{19}$, $\frac{9}{18}$, $\frac{4}{34}$ and $\frac{6}{16}$ have denominators which **do not** fit equally into 100.

Reach for a **calculator**: 'type in' these steps, and you will get a **decimal** as an answer:

$\frac{5}{19}$ Divide the top by the bottom: 5 ÷ 19 = 0.2631579
Round to 2 decimal places: **0.26** / approx. 26%

$\frac{9}{18}$ Divide the top by the bottom: 9 ÷ 18 = 0.5 (exactly 50%)

$\frac{4}{34}$ Divide the top by the bottom: 4 ÷ 34 = 0.1176470...
Round to 2 decimal places: **0.12** / approx. 12%

$\frac{6}{16}$ Divide the top by the bottom: 6 ÷ 16 = 0.375
Round to 2 decimal places: **0.38** / approx. 38%

TIP
This method works for **any** fraction. Try it with the questions on the previous page.

Try

Use a calculator to change these fractions into decimals and percentages:

$\frac{12}{30}$ $\qquad$ $\frac{3}{8}$ $\qquad$ $\frac{2}{15}$ $\qquad$ $\frac{5}{36}$ $\qquad$ $\frac{7}{8}$ $\qquad$ $\frac{4}{19}$ $\qquad$ $\frac{3}{12}$ $\qquad$ $\frac{5}{8}$

Now make up 2 more similar questions of your own!

Test

Use a calculator to change these fractions into decimals and percentages:

1. $\frac{6}{12}$

0.50 / 50%

2. $\frac{13}{19}$

0.6842105
0.68 (to 2 dp)
Approx. 68%

3. $\frac{3}{9}$

0.3333333
0.33 (to 2 dp)
Approx. 33%

4. $\frac{5}{9}$

0.5555555
0.56 (to 2 dp)
Approx. 56%

5. $\frac{7}{8}$

0.875
0.88 (to 2 dp)
Approx. 88%

6. $\frac{2}{9}$

0.2222222
0.22 (to 2 dp)
Approx. 22%

7. $\frac{3}{18}$

0.1666666
0.17 (to 2 dp)
Approx. 17%

8. $\frac{2}{8}$

0.25 / 25%

Now re-read the title I CAN statement; decide whether you have 'Achieved' or need to revisit.

Work Unit 3:
Shape and Geometry

Everyone's favourite part of the book (apart from **FLASH FACTS**☆™ perhaps!)

This unit is lighter than Units 1, 2 and 4 in that it is more visual and often regarded as more fun than all the number crunching stuff!

It is a good idea to do some of these pages **after** working on some of the number crunching activities, as they are light relief for the brain.

Unit 3 can be made more fun if you do the drawing / constructing shape activities on large paper (e.g. old wallpaper rolls or backs of posters) with colours, gel pens, felt tips or thin marker pens for variety.

You will also need a sharp pencil, a protractor, a ruler, a set square triangle (for right angles and constructing triangles) and a pair of compasses for drawing circles.

For tougher skills pages, where I have found in my experience that the student needs to have focussed, ultra high concentration levels, there is a fun 'warning': ① so you can ensure that the timing is just right to address the few concepts that are maybe a little harder to grasp, e.g. using a protractor.

For maximum learning impact and ownership of ability:
- Read out loud and discuss the 'I CAN / I KNOW' statement before starting a work page
- Repeat before the 8 'try' questions
- Repeat before the 8 'test' questions
- Finally, repeat once more at the end – then discuss whether you have 'Achieved' or need to revisit later

This helps the student be clear about the learning objective, tune in, focus and take ownership of the learning process. The student him/herself should tick the 'Achieved' box on the record chart.

The 'I CAN / I KNOW' Checklist for
Unit 3: Shape and Geometry

Page	I CAN / I KNOW	Tick & Date Visited	Achieved
145	I KNOW what 'parallel' means		❏
147	I KNOW what 'intersecting' means		❏
149	I KNOW what 'perpendicular' means		❏
151	I KNOW what 'congruent' means		❏
153	I KNOW what the 'D' means in 2D and 3D		❏
155	I KNOW prefixes for 2D shape names from 3 to 10 sides		❏
157	I KNOW the terms 'regular' and 'irregular' (2D shapes)		❏
159	I KNOW 3 important terms relating to properties of circles		❏
161	I KNOW how many degrees there are in circles and parts of circles		❏
163	I KNOW the names of 5 different angles		❏
(!) 165	I CAN use a protractor to measure angles less than 180 degrees		❏
(!) 167	I CAN use a protractor to construct angles less than 180 degrees		❏
169	I KNOW the names and basic properties of different types of triangles		❏
171	I CAN roughly sketch different types of triangles		❏
(!) 173	I CAN accurately construct different types of triangles		❏
175	I KNOW the names of 6 important quadrilaterals		❏

Page	I CAN / I KNOW	Tick & Date Visited	Achieved
177	I CAN roughly sketch 6 important quadrilaterals		❏
179	I KNOW how to find the perimeter of 2D shapes		❏
181	I KNOW the formula for finding the area of a square or rectangle		❏
(!) 183	I KNOW the formula for finding the area of triangles		❏
185	I KNOW 3 'transformation' terms		❏
187	I KNOW the names of 9 important 3D shapes		❏
189	I KNOW the terms 'faces', 'vertices' and 'edges' for 3D shapes		❏
191	I KNOW the special properties of prisms and pyramids		❏
193	I KNOW the formula for finding the volume of cubes and cuboids		❏

I KNOW what 'parallel' means

 We say lines are parallel if they are side by side, the same distance apart and will never meet, e.g. rail tracks.

... is an interesting and helpful word as it has a big clue in its spelling!
Spot the para**ll**el lines in the word.

Look at and discuss these examples of **parallel** lines and then draw some of your own on paper.

 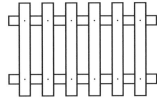 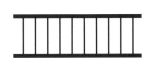

Using 2 rulers or pencils, make some pairs of **parallel lines** in the air.
Say, "**Parallel**" each time.

Use a **dictionary**: Look up the word 'parallel' to read the definition.

Try

Which of these pairs of lines are parallel?

Now make up 2 more similar questions of your own!

Test

Which of these pairs of lines are parallel?

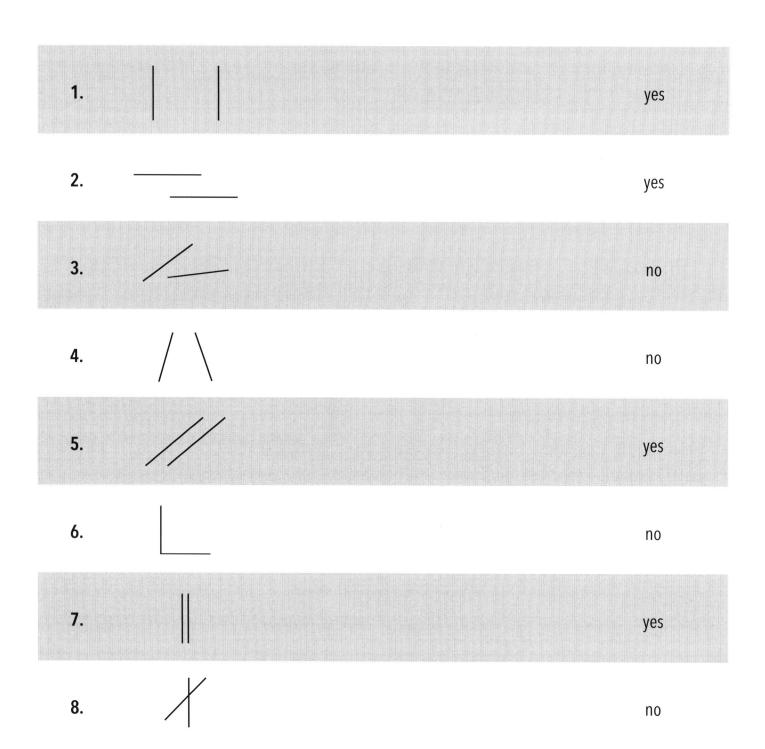

1. yes

2. yes

3. no

4. no

5. yes

6. no

7. yes

8. no

Now re-read the title I KNOW statement; decide whether you have 'Achieved' or need to revisit.

I KNOW what 'intersecting' means

We say lines are intersecting if they meet or cross each other.

The terms '**intersect**' or '**intersection**' are often used when referring to roads.

Look at and discuss these examples of **intersecting** lines and then draw some of your own on paper.

Using 2 rulers or pencils, make some pairs of **intersecting lines** in the air.
Say, "**Intersecting**" each time.

Use a dictionary: Look up the word 'intersect' or 'intersecting' to read the definition.

Try

Which of these pairs of lines are intersecting?

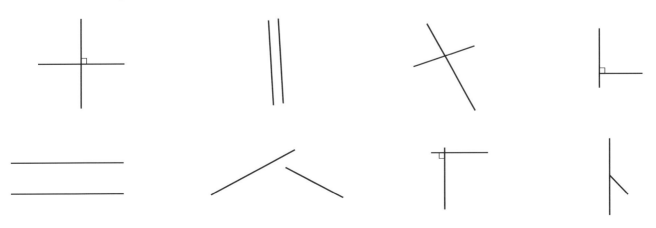

Now make up 2 more similar questions of your own!

Test

Which of these pairs of lines are intersecting?

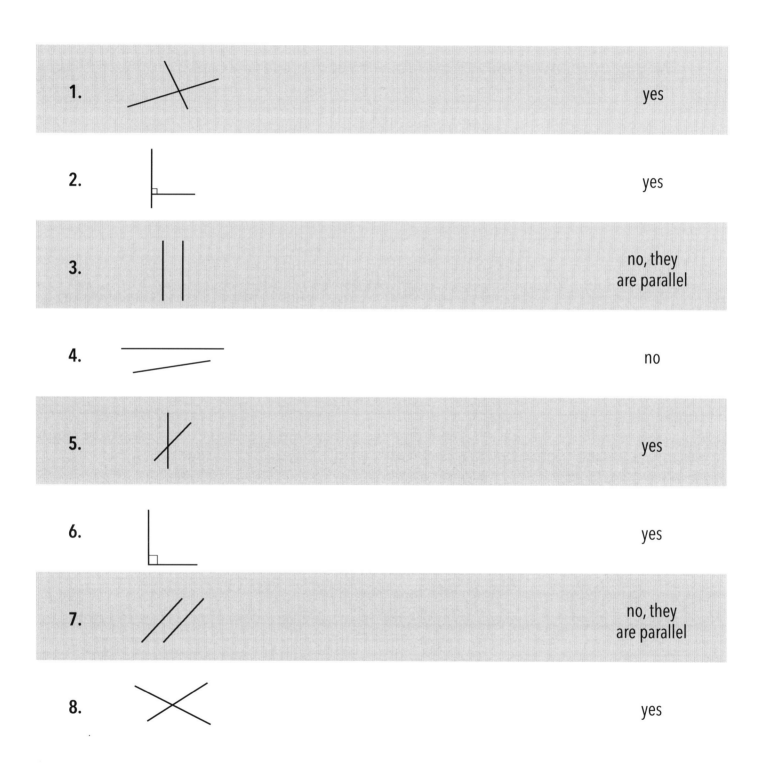

1.   yes

2.   yes

3.   no, they are parallel

4.   no

5.   yes

6.   yes

7.   no, they are parallel

8.   yes

Now re-read the title I KNOW statement; decide whether you have 'Achieved' or need to revisit.

I KNOW what 'perpendicular' means

We say lines are perpendicular if they intersect to form a right angle (90°), indicated by ⌐. Look out for the big right angle clue: the 'L' shape – but remember, the 'L' can be upside down or turned on its side.

'Per – pen – dic – u – lar' is a great word to roll around the tongue. Say it out loud a few times.

You can test whether 2 lines meet to form a right angle (90°) if you use a **set square** as a 'right angle' detector. If two lines intersect and form a perfect 'L' shape, then they make a right angle.

Look at and discuss these examples of **perpendicular** lines and then draw some of your own on paper.

Using 2 rulers or pencils, make some pairs of **perpendicular lines** in the air. Do some where the lines meet and do not cross, and some where they do cross. Say, "**Perpendicular**" each time.

Use a dictionary: Look up the word 'perpendicular' to read the definition.

Try

Identify which of these pairs of intersecting lines are perpendicular and mark on the right angle symbol ⌐

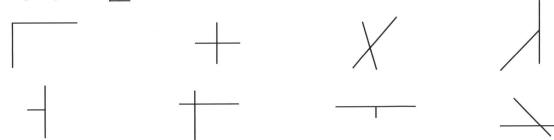

Now make up 2 more similar questions of your own!

Test

Identify which of these pairs of intersecting lines are perpendicular and mark on the right angle symbol.

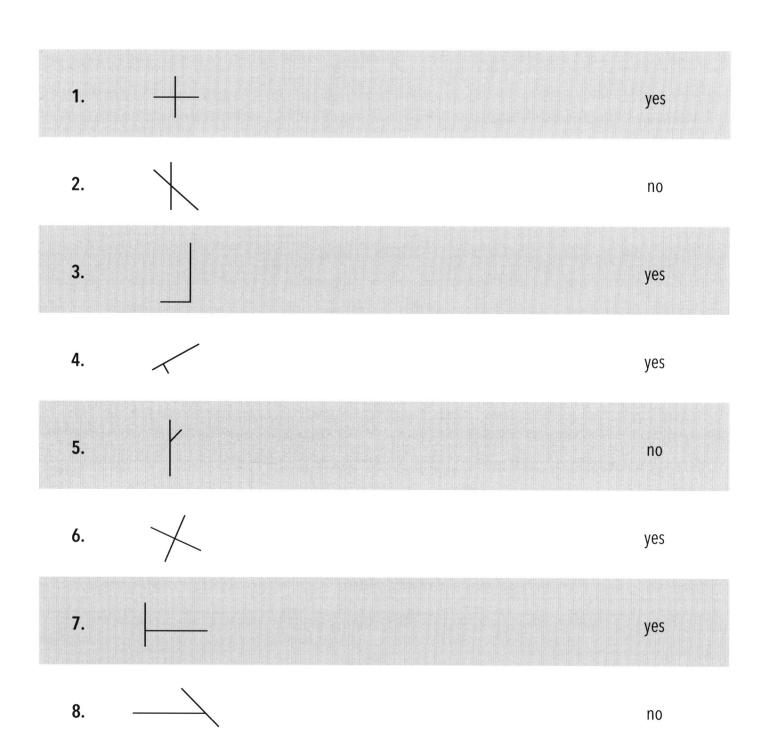

1.   yes

2.   no

3.   yes

4.   yes

5.   no

6.   yes

7.   yes

8.   no

Now re-read the title I KNOW statement; decide whether you have 'Achieved' or need to revisit.

I KNOW what 'congruent' means

 Congruent is a term which pops up in shape work.
It means 'identical in form', so if 2D shapes are congruent,
they are exactly the same.

However, the shapes may have been transformed (reflected or rotated – see later in this Unit), so it is not always easy to tell if they are exactly the same at first glance.

The best way to test for **congruency**, is to ask yourself if the image of the first shape would **sit perfectly on top** of the image of the second. You can test this out if you use tracing paper or kitchen parchment.

These examples show **congruent** pairs of 2D shapes:

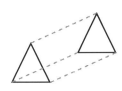  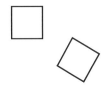 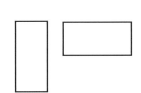

Use a **dictionary**: Look up the word 'congruent' to read the definition.

Try

Which of these pairs of shapes are congruent?

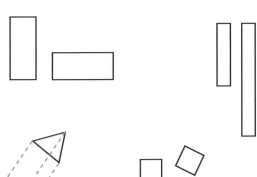  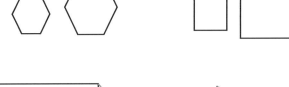 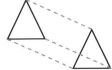

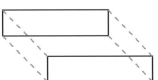

Now make up 2 more similar questions of your own!

Test

Are the shapes congruent?

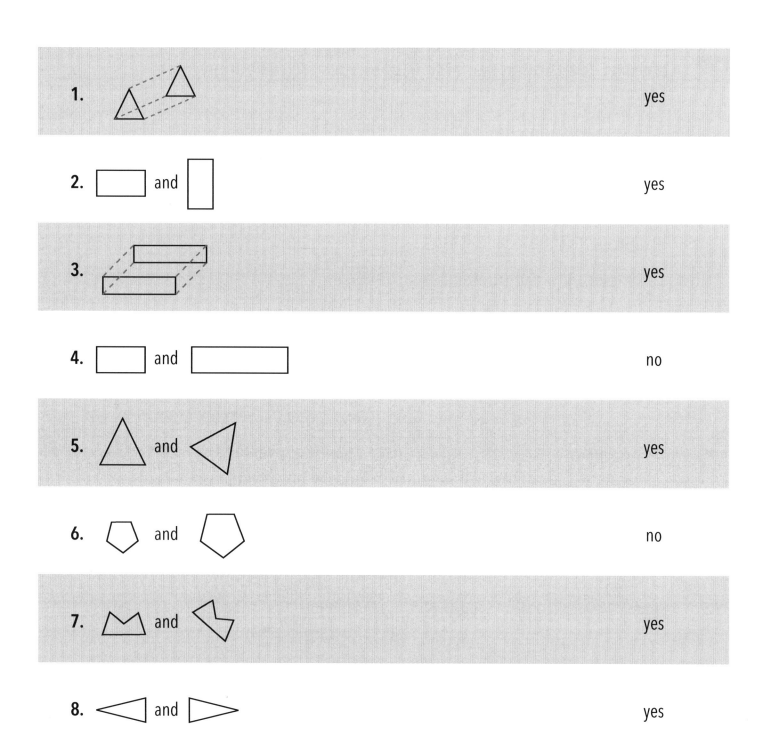

1.   yes

2.   and   yes

3.   yes

4.   and   no

5.   and   yes

6.   and   no

7.   and   yes

8.   and   yes

Now re-read the title I KNOW statement; decide whether you have 'Achieved' or need to revisit.

I KNOW what the 'D' means in 2D and 3D

We use the terms '2D' and '3D' when referring to shapes. 2D shapes are flat and 3D shapes are solid. Have you ever considered what the 'D' actually means?

It is referring to the **dimensions** of shapes:
2D means 2 dimensions (directions): **length and height**
3D means 3 dimensions (directions): **length, height and width**

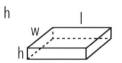

Our world is 3D. We have left and right / up and down / forwards and backwards.
Stand and 'do' these actions; you will 'feel' the dimensions:

dimension 1 is length (l)
across / left / right

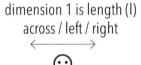

dimension 2 is height (h)
up / down

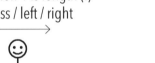

dimension 3 is width (w) or depth
forwards / backwards

Now find two or three small boxes to place in front of you. Trace your finger along the 3 dimensions **length (l), height (h) and width (w)**. Turn each box on its side, or twist it to change its position.

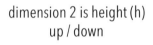 

Think about how and why the 3 dimensions (length, height and width) have now 'changed'. Look at a piece of furniture and do the same kind of thing with your eye.

Use a **dictionary**: Look up the word 'dimension' to read the definition.

Try

Look at these 2D and 3D shapes and maybe draw them.
Decide which word (length, height or width) best refers to each dimension in your mind:

Now make up 2 more similar questions of your own!

Test

Work through these questions and activities which require you to focus on 2D and 3D concepts:

1.	What does the 'D' mean in 2D and 3D?	dimensions
2.	In a 2D shape, what dimensions are there?	length and height
3.	Draw a square approximately 3 cm x 3 cm. Label the 2 dimensions length and height.	height / length
4.	In a 3D shape, what dimensions are there?	length, height and width
5.	A cuboid is a solid 3D shape. Can you copy it? Say what dimensions it has and label them.	h, l, w
6.	Look at and copy this shape: Is it 2D or 3D?	2D
7.	How many dimensions does this shape have?	it has 3 dimensions
8.	Draw some 3D boxes on paper. Label the dimensions.	e.g. h, l, w

Now re-read the title I KNOW statement; decide whether you have 'Achieved' or need to revisit.

I KNOW prefixes for 2D shape names from 3 to 10 sides

Prefixes can give us clever clues to numbers of sides on 2D shapes.

Look at these words:

A **tri**cycle: how many wheels does it have? **3**
A **tri**pod, e.g. to place a camera on: how many legs does it have? **3**
A **tri**angle: how many sides does it have? **3**

There are many other words which have the prefix 'tri' to give a clue of 'three'. Can you think of any? See if you can find some in a dictionary.

Look at how **prefixes** can give us clues for **numbers of sides on shapes**. Discuss the **prefix number clues** in the words in brackets, and maybe think of more examples.

3 sides: tri (3)   **tri**angle   (tripod, triathlon, tricycle)
4 sides: quad (4)   **quad**rilateral   (quadrangle, quad bike)
5 sides: pent (5)   **pent**agon   (Pentagon building in USA, pentathlon)
6 sides: hex (6)   **hex**agon   tip: '**hex**' as it has an '**x**' in it like the word '**six**'
7 sides: hept or sept (7) **hept**agon or **sept**agon (heptathlon, septet)
8 sides: oct (8)   **oct**agon   (octopus, octave)
9 sides: non (9)   **non**agon   tip: a bit crazy as sounds like 'none', but is 'nine'
10 sides: dec (10)   **dec**agon   (decimal, decade)

Try

Count the number of sides on these 2D shapes. Say the relevant prefix and then the 2D shape name:

Now make up 2 more similar questions of your own!

Test

Count the number of sides on these 2D shapes. Say the relevant prefix and then the 2D shape name:

1.   oct (8)

octagon

2.   quad (4)

quadrilateral

3.   pent (5)

pentagon

4.   hept or sept (7)

heptagon or septagon

5.   pent (5)

pentagon

6.   hex (6)

hexagon

7.   hex (6)

hexagon

8.   dec (10)

decagon

Now re-read the title I KNOW statement; decide whether you have 'Achieved' or need to revisit.

I KNOW the terms 'regular' and 'irregular' (2D shapes)

We talk about 2D shapes being either 'regular' or 'irregular'. Do you know what this means?

- If a 2D shape has sides all the same length and internal angles all the same size, it is 'regular'.
- If they are not all the same, it is 'irregular'.

Look at these 2 shapes. They are both **pentagons** because they have 5 sides.

The first one is **regular**: its sides are all the same length and its internal angles are all equal. The second one is **irregular**.

It is fun to make irregular shapes look like crazy birds or monsters. See how this **irregular hexagon** looks like a crazy bird:

Play around drawing **irregular** hexagons, octagons and decagons to make crazy creatures.

Try

Say whether these shapes are regular or irregular:

hexagon   octagon   pentagon   hexagon   decagon   pentagon   hexagon   octagon

Now make up 2 more similar questions of your own!

Test

Name these shapes and say whether they are regular or irregular:

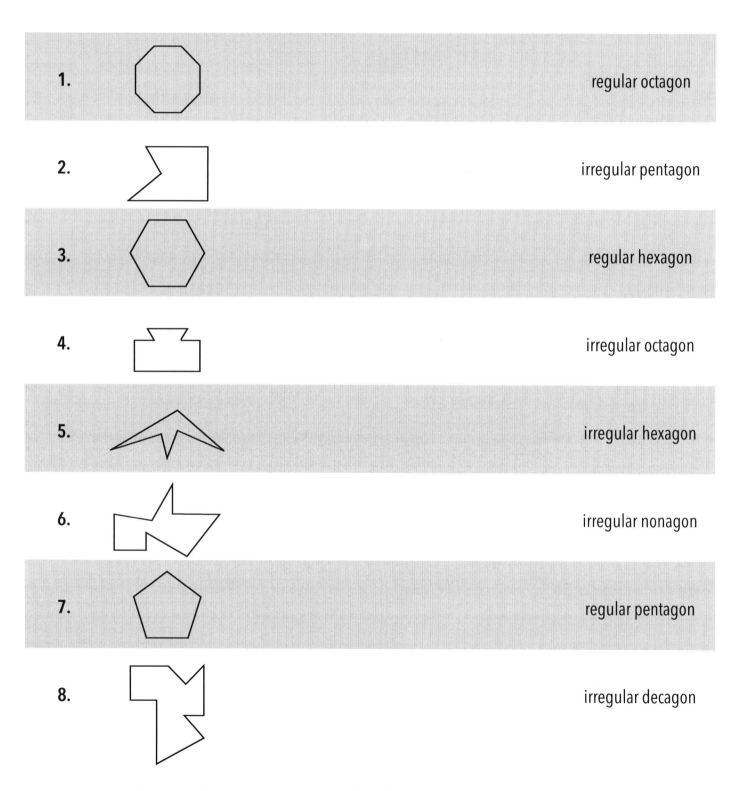

1. regular octagon

2. irregular pentagon

3. regular hexagon

4. irregular octagon

5. irregular hexagon

6. irregular nonagon

7. regular pentagon

8. irregular decagon

Now re-read the title I KNOW statement; decide whether you have 'Achieved' or need to revisit.

I KNOW 3 important terms relating to properties of circles

 There are 3 important terms you must know associated with circles: 'circumference', 'radius' and 'diameter'.

The first term begins with '**circ**' (as in **circ**le), so that is a helpful reminder:

- **circumference**: the distance around the circle

- **diameter**: the distance from one 'side' to the other (passing through the centre)

- **radius**: the distance from the centre to the 'side'

Of course, the **radius** is half the **diameter.** Stop and think about why this is true.

Using a pair of compasses, **draw different sized circles** on paper for fun. Label and measure the **diameter** and **radius** on some of them.

Pairs of compasses are **fun** to use, but quite **tricky** to handle at first, so you will need to play around with 'twizzling' to make circles or, alternatively, turning the paper whilst holding the compasses still. You can also create accurate circles by drawing around coins, small plates etc.

Use a dictionary: Look up the words 'circumference', 'radius' and 'diameter' to read the definitions.

Try

Work through these circumference / radius / diameter questions and activities:

What is this distance called?

Construct an accurate circle; label and measure the radius.

Using a pair of compasses, construct a circle with a diameter of 8 cm.

Write or say your own definiton for 'circumference'.

What is this distance called?

Construct an accurate circle; label and measure the diameter.

What is this distance called?

Construct an accurate circle and label the circumference.

Now make up 2 more similar questions of your own!

Test

Work through these circles-based questions and activities:

1.	Construct an accurate circle; label and measure the diameter.	e.g.

2.	Using a pair of compasses, construct a circle with a diameter of 6 cm. Think carefully about how wide you need to set the pair of compasses.	3 cm radius

3.	What does the dotted line represent?	radius

4.	If the radius of a circle was 10 cm, what would the diameter measure?	20 cm

5.	What does the dotted line represent?	circumference

6.	Finish the sentence: Radius is...	... the distance from the centre of a circle to the 'side'

7.	What does the dotted line represent?	diameter

8.	Finish the sentence: Diameter is...	... the distance from one 'side' of a circle to the other, passing through the centre

Now re-read the title I KNOW statement; decide whether you have 'Achieved' or need to revisit.

I KNOW how many degrees there are in circles and parts of circles

Have you heard the term, 'turn full circle'?
A full turn is 360 degrees (360°), so if you stand and face a point on a wall, and then turn a full 360°, you will be back where you started.

Look at this circle and see the **four quarter** markers. You need to learn these 4 important measurements / degrees.

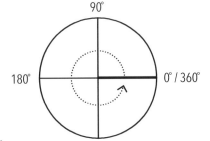

Make some circles and mark on the 4 important numbers to help you learn them.

Notice how the first four steps of the **9x Table** can help you remember the numbers:

1 x 9 = 9... add a zero **90°**   **2 x 9 = 18**... add a zero **180°**

3 x 9 = 27... add a zero **270°**   **4 x 9 = 36**... add a zero **360°**

Use, for example, a toothpick to 'pivot' from the centre point (lay it flat on the circle with one end at the centre point). Start at the zero / 360° mark each time and say the degree measurements out loud:

A quarter turn measures **90°**   **A half turn** measures **180°**
A three-quarter turn measures **270°**   **A full circle turn** measures **360°**

Try

How many degrees?

= _____° $\frac{3}{4}$ circle = _____°   = _____° $\frac{1}{4}$ circle = _____°

whole circle = _____°   = _____° $\frac{1}{2}$ circle = _____°   = _____°

Now make up 2 more similar questions of your own!

Test

How many degrees?

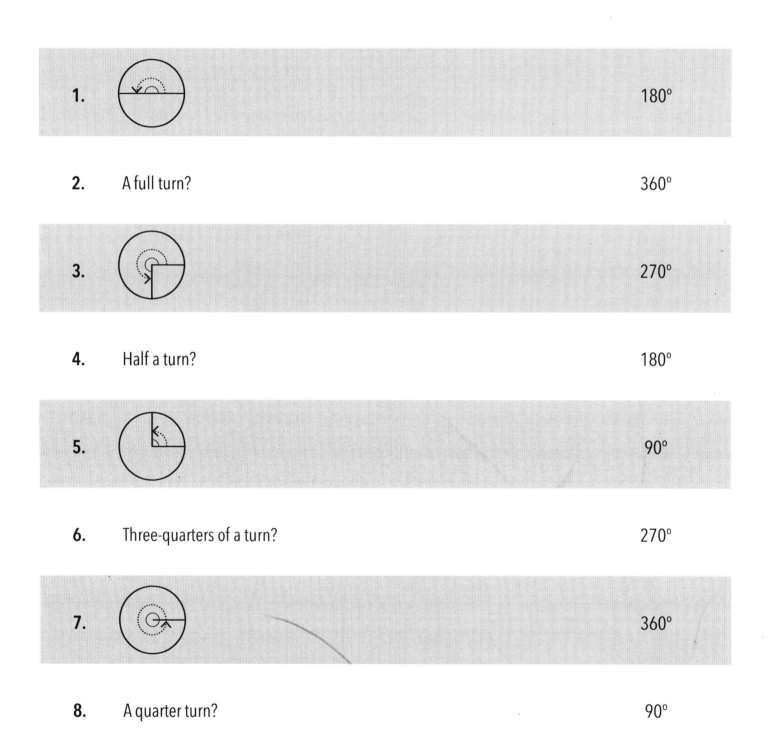

1. 180°

2. A full turn? 360°

3. 270°

4. Half a turn? 180°

5. 90°

6. Three-quarters of a turn? 270°

7. 360°

8. A quarter turn? 90°

Now re-read the title I KNOW statement; decide whether you have 'Achieved' or need to revisit.

I KNOW the names of 5 different angles

 An angle is the space between two intersecting lines. Angles are measured in degrees. We measure them using a protractor which we will learn how to use later. First, you need to know the names and properties of 5 types of angles.

right angle: exactly 90°
(tip: think 'L' shape – but it could be, for example, backwards or upside down)

acute angle: less than 90°
(tip: think 'cute' because it is small and sweet!)

obtuse angle: more than 90° but less than 180°

straight angle: exactly 180°
(tip: it makes a straight line)

reflex angle: more than 180°

Now pretend your outstretched arms are the two lines of an angle; starting with the **right angle** which is easy as it makes an 'L' shape, make all the **5 angles above** with your arms, saying their names as you go.

Apply the same idea using maybe 2 rulers, pencils or toothpicks on a table top.

Try

Identify these angles as acute, right, obtuse, straight or reflex:

Now make up 2 more similar questions of your own!

Test

Identify these angles as acute, right, obtuse, straight or reflex:

1. 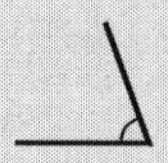 acute

2.  right

3.  obtuse

4.  acute

5.  straight

6.  right

7. 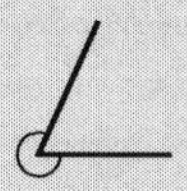 reflex

8.  obtuse

Now re-read the title I KNOW statement; decide whether you have 'Achieved' or need to revisit.

I CAN use a protractor to measure angles less than 180 degrees

At first, measuring angles using a protractor can be confusing, fiddly and tricky but once you get the hang of it, it is easy – think of it as a special type of 'ruler' to measure angles instead of straight lines.

Closely study your protractor.

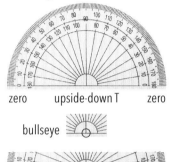

zero   upside-down T   zero

- Find the upside-down **T** positioned centrally between the two zeros. I call this point the **bullseye**!

bullseye

- Find the straight **baseline** which goes across the bottom from one zero to the other.

baseline

- There are **two number scales from 0 to 180**: count out loud, in tens, from 0 to 180 following the outside scale, then repeat in the opposite direction using the inside scale.

Look at this angle, and follow these steps to measure it:

1. Draw a small 'blob' on the point where the two lines intersect.
2. Place the **bullseye** on the 'blob'. Rotate the protractor until the **baseline** lays on top of a drawn line, as shown.
3. Say out loud, "Which zero have I used?" (point to the zero sitting on the **drawn** line).
4. Ask yourself if this is the inside or outside scale zero.
5. Follow round the scale which starts with that zero (10, 20, 30...).
6. Read off the angle measurement of the other (drawn) angle line.
7. If you followed the correct scale you should get **70°** for this angle. If you got 110°, you are reading the wrong scale.

TIP
Ask yourself what type of angle this is – see previous skill. It's an acute angle so it must have a value less than 90°

Try

Some of these angles are labelled with the wrong measurements – find them!

    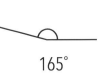

110°   15°   130°   40°   165°   80°   55°   120°

Now make up 2 more similar questions of your own!

Test

Measure these angles using a protractor:

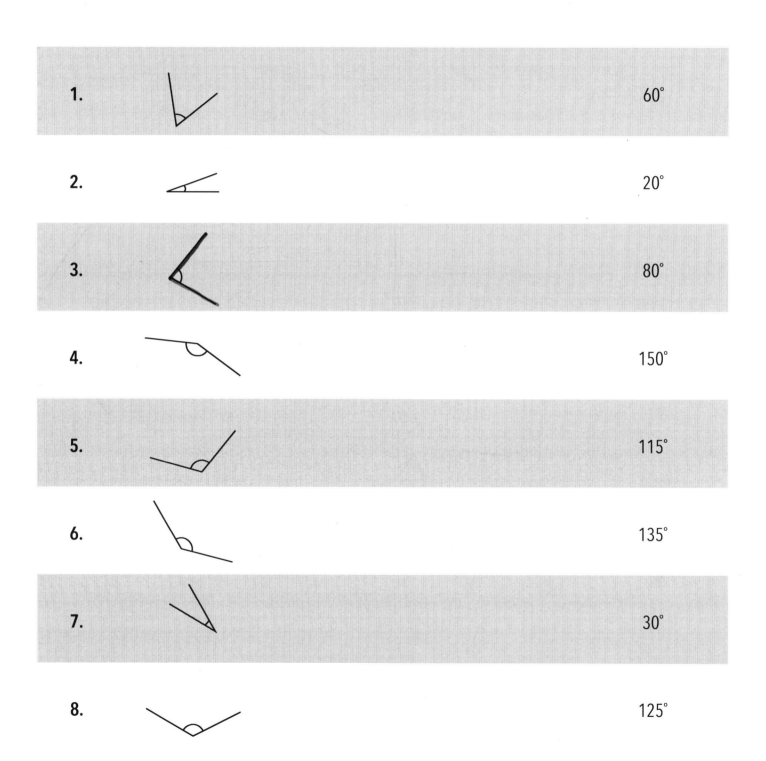

1.   60°

2.   20°

3.   80°

4.   150°

5.   115°

6.   135°

7.   30°

8.   125°

Now re-read the title I CAN statement; decide whether you have 'Achieved' or need to revisit.

I CAN use a protractor to construct angles less than 180 degrees

We can construct our own angles using a protractor.
You need a protractor, ruler, sharp pencil and some plain paper.

Follow the steps to **construct an angle measuring 70°**:

1. Draw a single line approximately 6 cm long, somewhere in the middle of the paper.
2. Draw a small 'blob' at one end (either end).
3. Place the protractor's **bullseye** on the 'blob' and line up your drawn line with the **baseline** of the protractor.

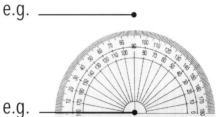

4. Say out loud, "Which zero have I used?" (point to the zero sitting on the drawn line).
5. Ask yourself if this is the inside or outside scale zero.
6. Follow round the scale which starts with that zero to **70°** and put a pencil mark (just outside the protractor edge) on the paper to show where **70°** is.
7. Now remove the protractor, and draw in the second ruler line from the blob to the pencil mark to make your **70°** angle.

8. Now double check it by re-measuring your angle with the protractor to check it measures 70°. Label it 70°.

Try

Construct angles measuring:

50°   20°   80°   175°   115°   135°   30°   40°

Now make up 2 more similar questions of your own!

Test

Construct angles measuring:

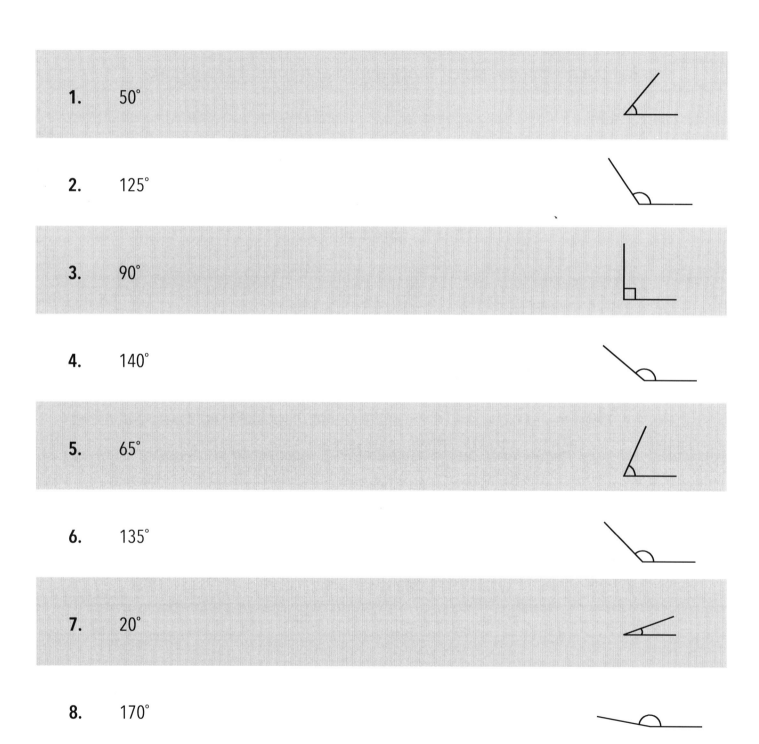

1.   50°

2.   125°

3.   90°

4.   140°

5.   65°

6.   135°

7.   20°

8.   170°

Now re-read the title I CAN statement; decide whether you have 'Achieved' or need to revisit.

I KNOW the names and basic properties of different types of triangles

Triangles all have three sides and angles of course, but there are different types of triangles.

Triangles: please RISE and show yourselves!

You need to be able to identify them and know their special properties.

	Name	Hint	Properties	Example
R	Right angled	It has a right angle. Simple. It will be a right angled isosceles or a right angled scalene depending on lengths of sides.	Easy to identify – if you spot a right angle, you have its name.	
I	Isosceles	I say, "I love sausages" to remind me. I know it's silly, but it helps me remember.	Two sides the same length and one different, e.g. a witch hat.	
S	Scalene	I call this the 'Three Bears Triangle' because it has a baby bear side, mummy bear side and a daddy bear side (small, medium and large).	Three sides all different lengths – short, medium and long.	
E	Equilateral	Look how the word starts. It sounds like 'equal' – all sides are equal lengths.	Three sides all the same length and three internal angles all the same.	

FACT
The three internal angles of all triangles always add up to 180˚. Check it out.

Try

Name each type of triangle and point out the properties:

Now make up 2 more similar questions of your own!

Test

Name each type of triangle:

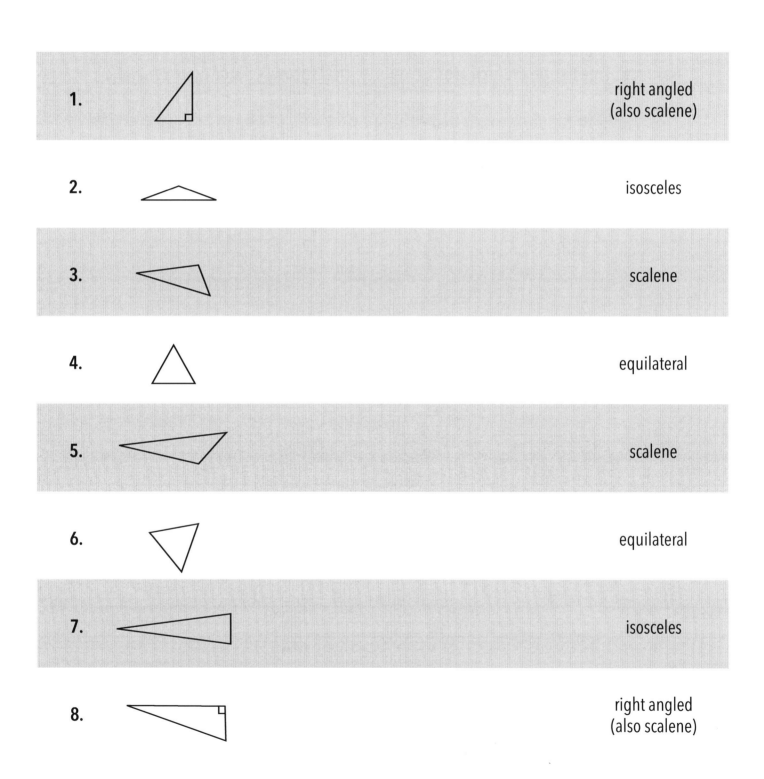

1.   right angled (also scalene)

2.   isosceles

3.   scalene

4.   equilateral

5.   scalene

6.   equilateral

7.   isosceles

8.   right angled (also scalene)

Now re-read the title I KNOW statement; decide whether you have 'Achieved' or need to revisit.

I CAN roughly sketch different types of triangles

Roughly sketching different types of triangles is fun, and for this exercise they don't have to be perfect. Draw lots of different types of triangles to get used to how they look.

Work freehand if you want, or use a ruler; maybe add colour or make fun pictures.

R **Right angled**: easy!
Start by playing around with the letter 'L' which can be stretched out of proportion, tiny, huge or rotated.
When you add the final side, it is complete.
It will be a right angled isosceles or a right angled scalene depending on lengths of sides.

e.g.

I **Isosceles**:
2 sides the same length and one a different length.
Draw a witch hat or an ice cream cone.
Remember though, it could have 2 shorter sides and one longer side.

e.g.

S **Scalene**:
3 sides all different lengths – small, medium and large (baby, mummy and daddy bears).
Draw a short horizontal line for the first side.
Draw a second side which is longer, heading off on a **sharp** tilt right or left.
Draw the final side. The 3 sides should be different lengths.
Measure them to check!

e.g.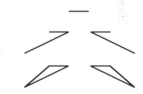

E **Equilateral**:
3 sides all the same length and 3 internal angles all the same.
Play around until your 3 sides are all the same length.

e.g.

Try

On fresh paper, roughly sketch different sized versions of the triangles .
Name and label them.

Test

Sketch these triangles using approximate measurements:

1. Sketch a scalene triangle with one side 2 cm long.   e.g.

2. Sketch a right angled triangle with one side 6 cm long. Mark on the right angle.   e.g.

3. Sketch an equilateral triangle with sides of 5 cm.   e.g.

4. Sketch an isosceles triangle with a base of 4 cm.   e.g.

5. Sketch a scalene triangle with one side 3 cm long.   e.g.

6. Sketch a right angled triangle with one side 14 cm long. Mark on the right angle.   e.g.

7. Sketch an isosceles triangle with a base of 12 cm.   e.g.

8. Sketch an equilateral triangle with sides of 7 cm.   e.g.

Now re-read the title I CAN statement; decide whether you have 'Achieved' or need to revisit.

I CAN accurately construct different types of triangles

For geometry work, constructing different types of triangles sometimes needs you to be accurate and specific with measurements of lines and angles to produce neat 'technical' drawings.

You need a set square, a protractor, a ruler and a sharp pencil. Follow the steps to construct these triangles, and label the side lengths (ruler) and internal angles (protractor). Remember, **internal angles** of all triangles add up to **180°**.

R **Right angled**:
1. Use the set square to make your starting 90° L, making two sides.
2. Now draw in the third side.

e.g.

I **Isosceles**:
1. Draw a horizontal base 4 cm; place a dot at the half way point.
2. Make an erasable, faint vertical guide line from the dot, using the right angled corner of a set square so the guide line is at right angles (perpendicular) to the base.
3. Draw in the two equal length sides (not the same length as the base) making them meet at the guideline.

e.g.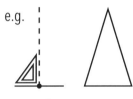

S **Scalene**:
1. Draw a 3 cm horizontal line.
2. Draw a second side 8 cm, heading off on a sharp tilt right or left.
3. Draw the final side. The 3 sides should be different lengths. Measure them to check!

e.g.

E **Equilateral**:
1. Draw a horizontal base 6 cm; place a dot at the half way point.
2. Make an erasable, faint vertical guide line in the middle, using a set square so the guide line is at right angles (perpendicular) to the base.
3. Draw in the remaining two 6 cm sides making them meet at the guideline; check all sides are equal.

e.g.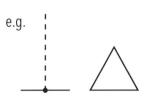

Try

Accurately construct different sized versions of the triangles.
Name and label them.

Test

Accurately construct these triangles:

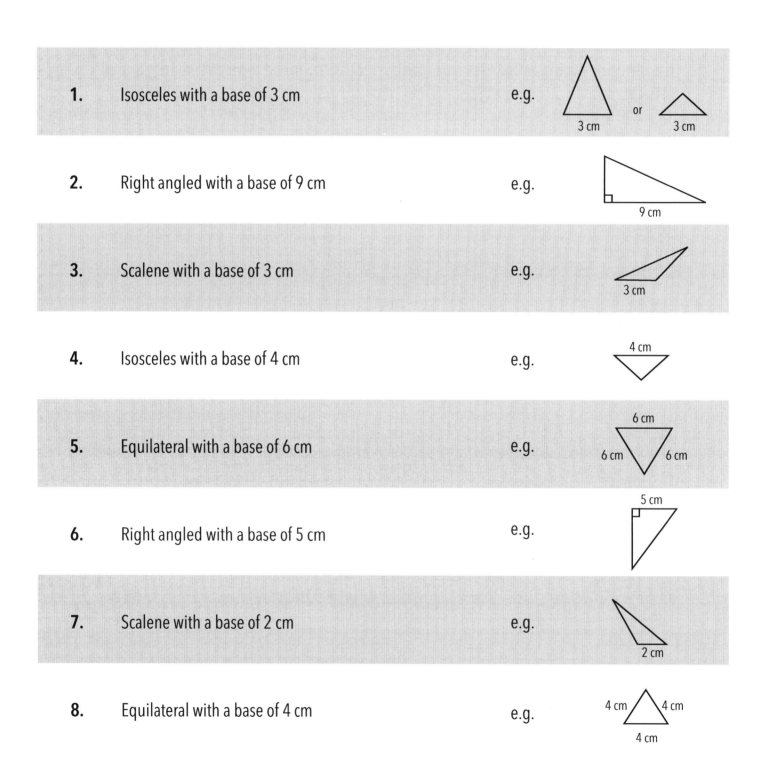

1.	Isosceles with a base of 3 cm	e.g.
2.	Right angled with a base of 9 cm	e.g.
3.	Scalene with a base of 3 cm	e.g.
4.	Isosceles with a base of 4 cm	e.g.
5.	Equilateral with a base of 6 cm	e.g.
6.	Right angled with a base of 5 cm	e.g.
7.	Scalene with a base of 2 cm	e.g.
8.	Equilateral with a base of 4 cm	e.g.

Now re-read the title I CAN statement; decide whether you have 'Achieved' or need to revisit.

I KNOW the names of 6 important quadrilaterals

 Quadrilaterals are 2D shapes with four sides. Remember how the prefix 'quad' can give us a clue for 'four', e.g. quad bike (4 wheels). Check you know the names (and spellings) of these 6 important quadrilaterals:

square　　　　this **quad**rilateral is usually the first one we learn as young students

rectangle　　another well known **quad**rilateral

trapezium　　tip:　looks like it could catch and '**trap**' a tiny mouse

kite　　　　　tip: 2 equal length shorter sides, 2 equal length longer sides

rhombus　　　tip: all the sides are the same length

parallelogram　tip: has 2 sets of **parallel** lines

Try

Name the quadrilaterals:

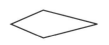 

Now make up 2 more similar questions of your own!

Test

Name the quadrilaterals:

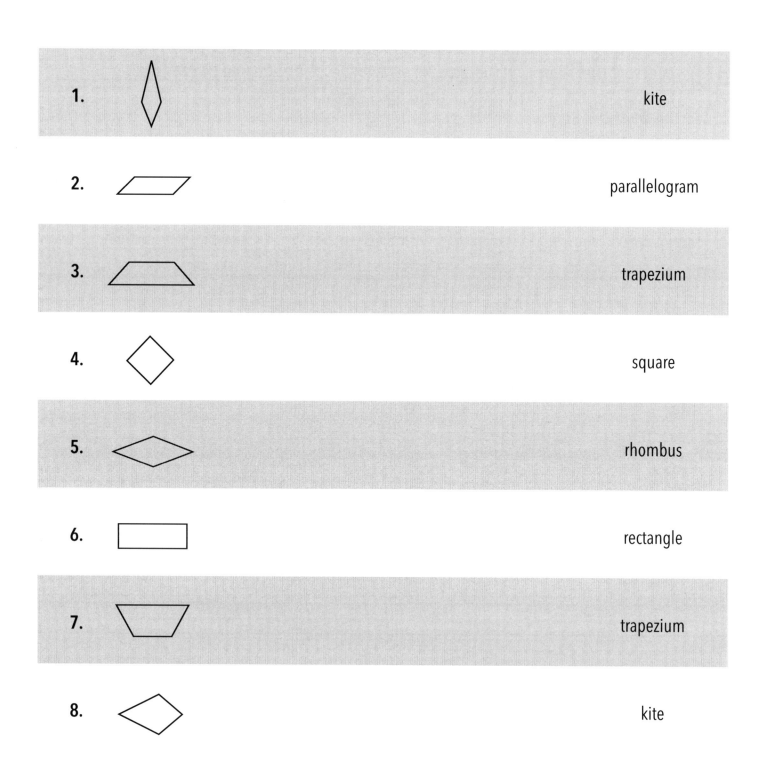

1.   kite

2.   parallelogram

3.   trapezium

4.   square

5.   rhombus

6.   rectangle

7.   trapezium

8.   kite

Now re-read the title I KNOW statement; decide whether you have 'Achieved' or need to revisit.

I CAN roughly sketch 6 important quadrilaterals

Sketching these six quadrilaterals is fun, and they don't have to be perfect. Trace over and then copy (freehand) these quadrilaterals:

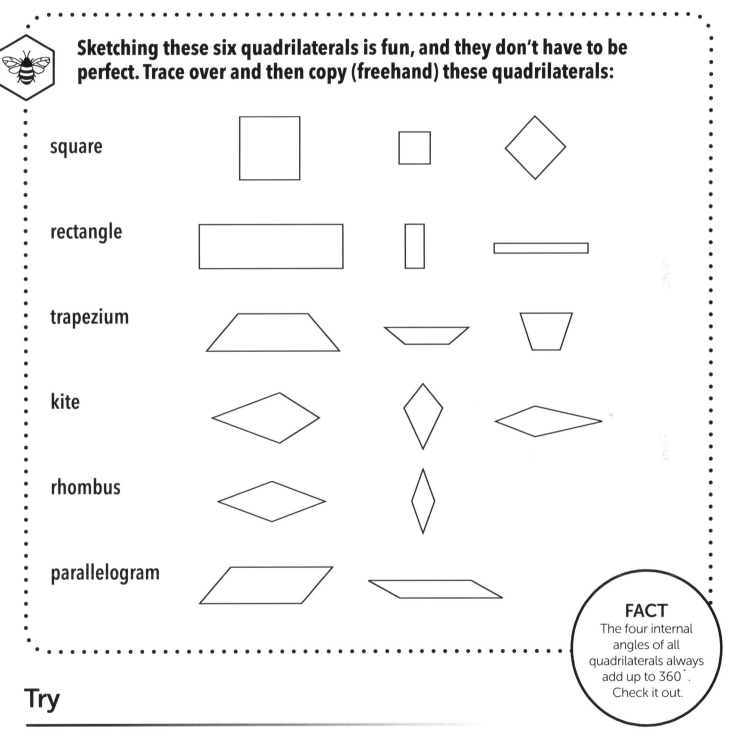

square

rectangle

trapezium

kite

rhombus

parallelogram

FACT
The four internal angles of all quadrilaterals always add up to 360°. Check it out.

Try

On fresh paper, roughly sketch 2 different sized versions of these 6 quadrilaterals. Then maybe do some more using a ruler so they begin to look more professional. Name and label them.

Test

Construct these quadrilaterals as accurately as you can, using a ruler.

1. A parallelogram with its longer sides measuring 8 cm e.g.

2. A rhombus with all sides measuring 6 cm e.g.

3. A kite with its longer sides measuring 5 cm e.g.

4. A trapezium with one of its sides measuring 4 cm e.g.

5. A rectangle with its two longer sides measuring 7 cm e.g.

6. A square with all sides measuring 3 cm e.g.

7. A parallelogram with its longer sides measuring 7 cm e.g.

8. A rhombus with all sides measuring 12 cm e.g.

Now re-read the title I CAN statement; decide whether you have 'Achieved' or need to revisit.

I KNOW how to find the perimeter of 2D shapes

**Perimeter is the distance around a 2D shape.
Think of a worm crawling round from a starting place on the outside of a 2D shape, and getting back to where he started.**

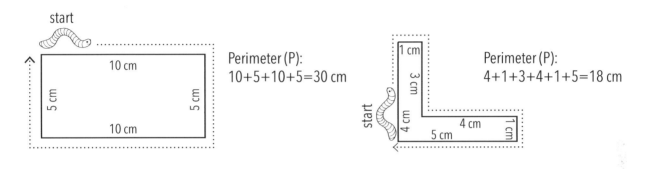

start

Perimeter (P):
10+5+10+5=30 cm

Perimeter (P):
4+1+3+4+1+5=18 cm

start

Perimeter (P):
5+5+5+5=20 cm

Perimeter (P):
20+20+20+20+20+20=120 m

Use a dictionary: Look up the word 'perimeter' to read the definition.

Try

Find the perimeter of these 2D shapes:

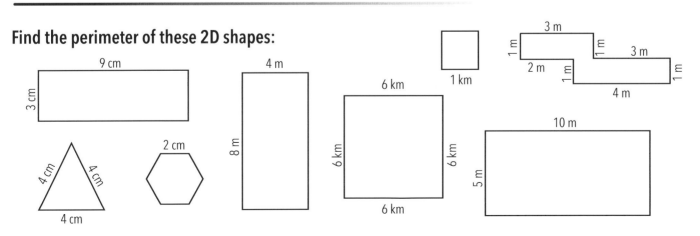

Now make up 2 more similar questions of your own!

Test

Find the perimeter of these 2D shapes:

1.
6 cm
1 cm

$6 + 1 + 6 + 1 = 14$ cm

2.
10 cm
4 cm
10 cm

$4 + 10 + 10 = 24$ cm

3.
10 m

$10 + 10 + 10 + 10 = 40$ m

4.
3 cm

$3 + 3 + 3 + 3 + 3 = 15$ cm

5.
3 m
1 m
1 m
3 m
2 m
1 m
1 m
4 m

$3 + 1 + 3 + 1 + 4$
$+ 1 + 2 + 1 = 16$ m

6.
3 cm
2 cm

$2 + 3 + 2 + 3 = 10$ cm

7.
2 cm
6 cm

$6 + 2 + 6 + 2 = 16$ cm

8.
4 m

$4 + 4 + 4 = 12$ m

Now re-read the title I KNOW statement; decide whether you have 'Achieved' or need to revisit.

I KNOW the formula for finding the area of a square or rectangle

Area is how much space there is on a flat surface, or inside a perimeter. To find the area of squares or rectangles, there is a formula (a special trick): multiply length by height (l x h).

Some examples:

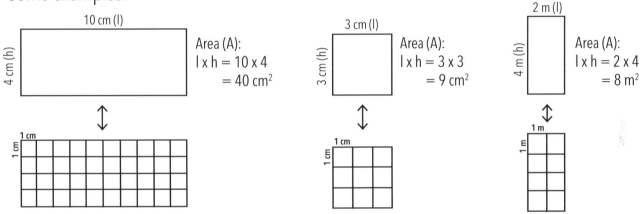

10 cm (l), 4 cm (h)
Area (A):
l x h = 10 x 4
= 40 cm²

3 cm (l), 3 cm (h)
Area (A):
l x h = 3 x 3
= 9 cm²

2 m (l), 4 m (h)
Area (A):
l x h = 2 x 4
= 8 m²

Area can be measured in square millimetres (**mm²**), square centimetres (**cm²**), square metres (**m²**), square kilometres (**km²**), or some other 'square' units.

If an irregular shape can be 'cut up' to make neat squares or rectangles, draw dividing lines, work out the area of the separate parts, then add them together for the total area.

e.g.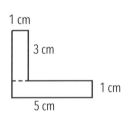

Area (A):
1 x 3 = 3 cm²
5 x 1 = 5 cm²
A = 3 + 5 = 8 cm²

TIP
Notice the connection between '2D', '2 dimensions' and the 'tiny number 2' e.g. cm²

Try

Find the areas:

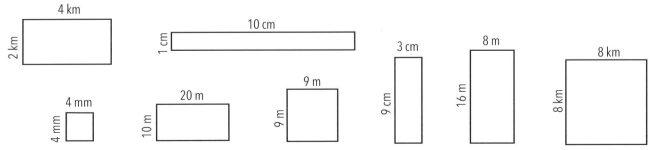

Now make up 2 more similar questions of your own!

Test

Find the areas:

1.

8 cm

2 cm

$8 \times 2 = 16 \text{ cm}^2$

2.

20 km

5 km

$20 \times 5 = 100 \text{ km}^2$

3.

5 m

$5 \times 5 = 25 \text{ m}^2$

4. A notice board is 2 m long and 1 m high. What is its area?

$2 \times 1 = 2 \text{ m}^2$

5.

5 cm

10 cm

$5 \times 10 = 50 \text{ cm}^2$

6.

4 m

3 m

$4 \times 3 = 12 \text{ m}^2$

7.

2 cm

3 cm

$2 \times 3 = 6 \text{ cm}^2$

8. One side of a square poster measures 10 cm. What is its area?

$10 \times 10 = 100 \text{ cm}^2$

Now re-read the title I KNOW statement; decide whether you have 'Achieved' or need to revisit.

I KNOW the formula for finding the area of triangles

Area is how much space there is on a flat surface, or inside a perimeter. To find the area of triangles, there is a formula (a special trick): multiply half the base by the height ($\frac{1}{2}$ b x h).
Note: some triangles might not have the 'base' in the most obvious place!

Some examples:

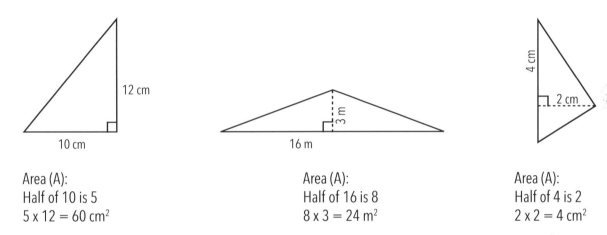

Area (A):
Half of 10 is 5
5 x 12 = 60 cm²

Area (A):
Half of 16 is 8
8 x 3 = 24 m²

Area (A):
Half of 4 is 2
2 x 2 = 4 cm²

When finding the area of a triangle, all you need to know are the **base** and **height measurements** (the height is sometimes given as an extra piece of information).

Sometimes the lengths of ALL sides are shown – but all you need to look at are the **base and the height** – ignore the other measurements.

TIP
Notice the connection between '2D', '2 dimensions' and the 'tiny number 2' e.g. cm²

Try

Find the area of these triangles:

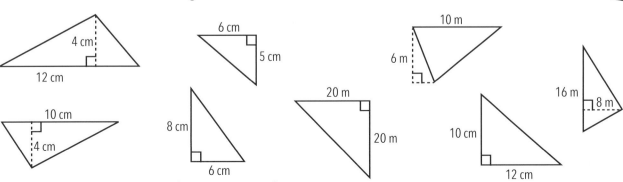

Now make up 2 more similar questions of your own!

Test

Find the area of these triangles:

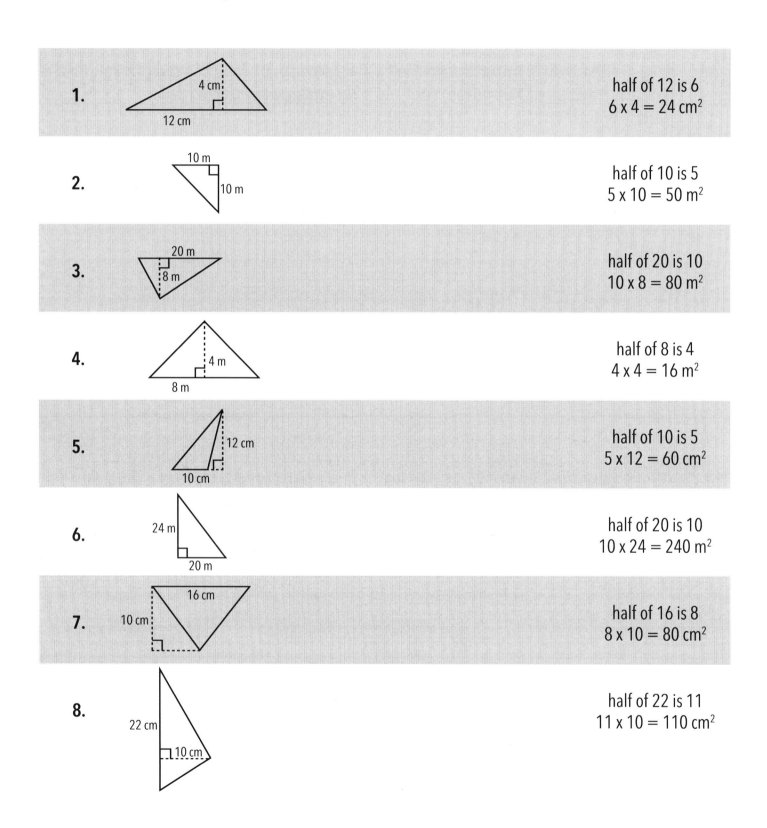

1.
half of 12 is 6
6 x 4 = 24 cm²

4 cm
12 cm

2.
half of 10 is 5
5 x 10 = 50 m²

10 m
10 m

3.
half of 20 is 10
10 x 8 = 80 m²

20 m
8 m

4.
half of 8 is 4
4 x 4 = 16 m²

4 m
8 m

5.
half of 10 is 5
5 x 12 = 60 cm²

12 cm
10 cm

6.
half of 20 is 10
10 x 24 = 240 m²

24 m
20 m

7.
half of 16 is 8
8 x 10 = 80 cm²

16 cm
10 cm

8.
half of 22 is 11
11 x 10 = 110 cm²

22 cm
10 cm

Now re-read the title I KNOW statement; decide whether you have 'Achieved' or need to revisit.

I KNOW 3 'transformation' terms

To 'transform' means to 'change'. Basic transformation in maths is when a 2D shape is moved or changed to another position. Here are three basic transformations to learn about:

translation / translate (slide): an image slides up, down, left, right or diagonally

e.g.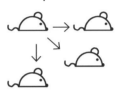

rotation / rotate (turn): an image turns about a point of rotation; the turn is often 90° or 180° (clockwise or anti-clockwise) but could be any angle up to 360°

e.g.

reflection / reflect (flip): an image flips across a mirror line

e.g.

TIP
When working on transformations, reach for tracing paper or kitchen parchment – this will help you, as you can trace the starting image before you figure out what has been 'done' to it.

Try

Fit the words into the spaces:

change   flip   turn   slide   reflection   rotation   translation   mirror line

Translate means _____

This image shows a _____

Reflect means _____

This image shows a _____ **j | i**

Transform means _____

We flip an image over a _____

Rotate means _____

This image shows a _____

Now make up 2 more similar questions of your own!

Test

Fit the words into the spaces:

flip   slide   change   turn   reflection   translation   transformations   rotation

1.   This image shows a _____   rotation

2.   To translate means to _____   slide

3.   The term 'transform' means _____   change

4.   Reflection means to _____   flip

5.   Rotation is to _____   turn

6.   This image shows a _____   reflection

7.   This image shows a _____   translation

8.   Rotation, reflection and translation are all types of _____   transformations

Now re-read the title I **KNOW** statement; decide whether you have 'Achieved' or need to revisit.

I KNOW the names of 9 important 3D shapes

You need to know the names of these 9 important 3D shapes:

cube

sphere

cuboid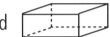

square pyramid

triangular prism

rectangular pyramid

cylinder   cone

triangular pyramid

Using tracing paper (or kitchen parchment) trace over all the 3D shapes, and then try to draw your own versions on fresh paper using freehand (or maybe using a ruler where appropriate). As well as having fun drawing them, remember to say the names of them and label them; this might be a good opportunity to make sure you can **spell them correctly**.

TIP
One common mistake many students make is that they call a **cube** a square, and a **cuboid** a rectangle. Make sure you know the differences: squares and rectangles are 2D flat shapes, cubes and cuboids are **3D solid shapes**.

Try

Name the 3D shapes:

      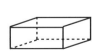

Test

Name the 3D shapes:

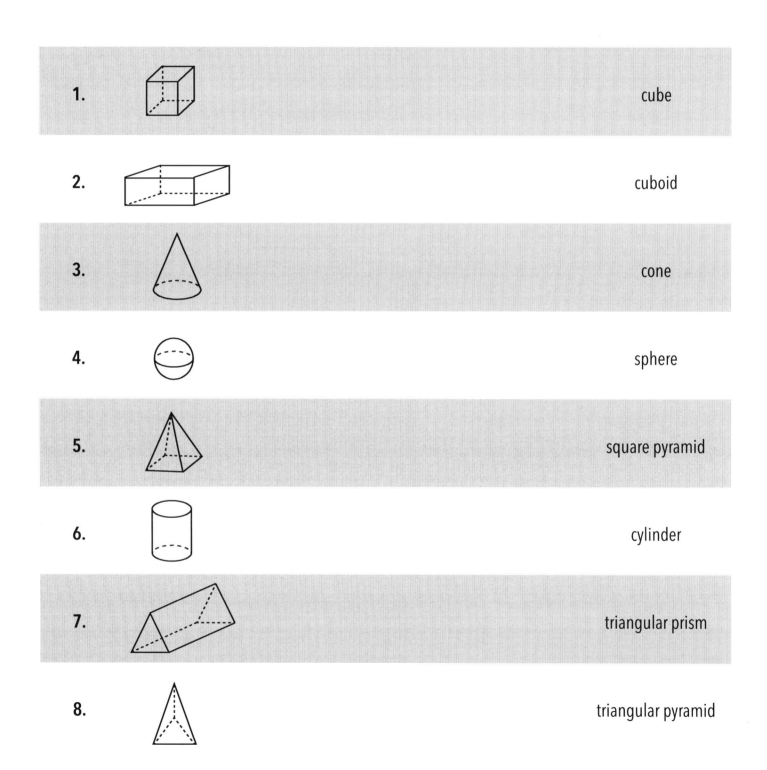

1.		cube
2.		cuboid
3.		cone
4.		sphere
5.		square pyramid
6.		cylinder
7.		triangular prism
8.		triangular pyramid

Now re-read the title I KNOW statement; decide whether you have 'Achieved' or need to revisit.

I KNOW the terms 'faces', 'vertices' and 'edges' for 3D shapes

Three dimensional (3D) shapes have faces, vertices and edges. You need to use these terms when referring to their properties.

Find a real **cuboid** (maybe a box of tissues) or a **cube** to hold, touch and see.

faces: lay out your hand flat, and move your palm across every **face** (flat surface).

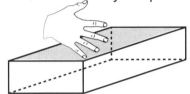

vertices: point to every corner of the box. Say "OUCH!" each time. These are the **vertices**. Note: the singular of vertices is **'vertex'**.

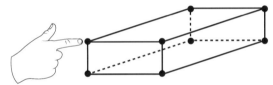

edges: run your finger along the 'folds' of the box. These are the **edges**.

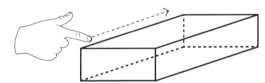

Now count up how many **faces**, **vertices** and **edges** your cuboid has. You should have counted 6 faces, 8 vertices and 12 edges.

Now find a **different type of 3D shape** and do the same.

Try

Say whether faces, vertices or edges are highlighted on these 3D shapes.

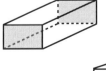 

Now make up 2 more similar questions of your own!

Test

Say whether faces, vertices or edges are highlighted on these 3D shapes.

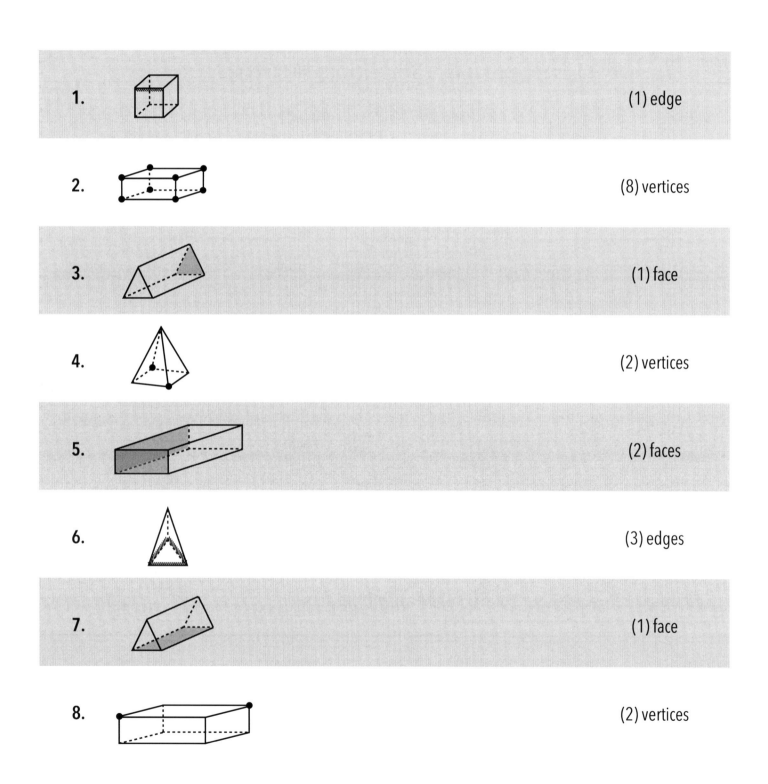

1.   (1) edge

2.   (8) vertices

3.   (1) face

4.   (2) vertices

5.   (2) faces

6.   (3) edges

7.   (1) face

8.   (2) vertices

Now re-read the title I KNOW statement; decide whether you have 'Achieved' or need to revisit.

I KNOW the special properties of prisms and pyramids

Prisms and pyramids are very interesting 3D shapes. If you look carefully at examples of them, you will find clues for their names.

A **prism** is a 3D solid whose **two ends are congruent**.
Look at the 2D shape at either end to get a clue for its name.

triangular prism

rectangular prism (cuboid)

pentagonal prism

Many congruent pairs of 2D shapes could be at the ends. To identify the name of the **prism**, look to see which 2D shape is at **both ends** and say out loud: "a _____ prism".

A **pyramid** is a 3D solid which has a 2D shape as its **base**, and it also goes to a **point**.

Square pyramid

rectangular pyramid

triangular pyramid

hexagonal pyramid

Many different 2D shapes could be the base. To identify the name of the pyramid, look to see which 2D shape is on the base and say out loud: "a _____ pyramid".

Have a go at drawing some of the above **prisms** and **pyramids** on paper. Maybe add colour. For the prisms, start with the two congruent ends. For the pyramids, start with the base, then place a dot above for the point.

Try

Name these solids:

Test

Name these solids:

1.		triangular prism
2.		square pyramid
3.		triangular pyramid
4.		rectangular prism (cuboid)
5.		rectangular pyramid
6.		hexagonal prism
7.		octagonal prism
8.		pentagonal pyramid

Now re-read the title I KNOW statement; decide whether you have 'Achieved' or need to revisit.

I KNOW the formula for finding the volume of cubes and cuboids

Volume is how much three dimensional space there is inside a 3D shape. To find the volume of cubes or cuboids, there is a formula (a special trick): multiply length by height by width (l x h x w).

Some examples:

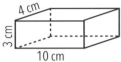

Volume (V):
l x h x w = 10 x 3 x 4
= 120 cm³

cuboid

Volume (V):
l x h x w = 4 x 4 x 4
= 64 cm³

cube

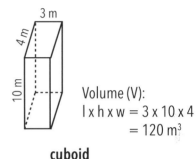

Volume (V):
l x h x w = 3 x 10 x 4
= 120 m³

cuboid

Volume can be measured in cubic centimetres (**cm³**), cubic metres (**m³**) or some other 'cubic' units. It is about how many **cubic units** would fit inside a solid if you took off the 'lid' and stacked them in.

cubic centimetres (cm³): picture the little centimetre cubes you used in early maths lessons to learn about ones, tens and hundreds.

cubic metres (m³): picture how you would make a huge 3D cube out of metre sticks and sticky tape, or maybe picture the 'box' a washing machine comes in. You could fit inside it! These measurements would be used to refer to volume of large containers, e.g. a shipping container or a warehouse.

It depends which way up a solid is positioned as to which dimension (l, h, w) is which; it doesn't matter – just multiply all three (sometimes 'width' is called 'depth').

TIP
Notice the connection between '3D', '3 dimensions' and the 'tiny number 3' e.g. cm³

Try

Find the volume of these 3D shapes:

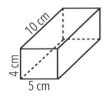
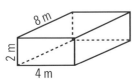

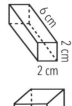

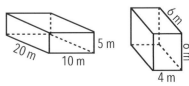

Now make up 2 more similar questions of your own!

Test

Work through these volume questions:

1.	What is the formula for finding the volume of a cube or cuboid?	l x h x w
2.	A cube has a height of 2 cm. What is its volume?	$2 \times 2 \times 2 = 8 \text{ cm}^3$
3.	This box is a cuboid. What is its volume?	$2 \times 3 \times 10 = 60 \text{ cm}^3$
4.	A box has dimensions of 5 cm, 3 cm and 4 cm. Find its volume.	$5 \times 3 \times 4 = 60 \text{ cm}^3$
5.	Here is a large storage box. What is its volume?	$2 \times 4 \times 2 = 16 \text{ m}^3$
6.	Find the volume:	$10 \times 10 \times 3 = 300 \text{ cm}^3$
7.	What is the volume of a centimetre cube?	$1 \times 1 \times 1 = 1 \text{ cm}^3$
8.	Find the volume of a box 4 m x 5 m x 2 m?	$4 \times 5 \times 2 = 40 \text{ m}^3$

Now re-read the title I KNOW statement; decide whether you have 'Achieved' or need to revisit.

Work Unit 4:
Times Tables, Exponents and Prime Numbers

To 'bridge the maths gap', this Unit is central.

It is a vital bridge to accessing the maths curriculum from now on.

A student who knows all Times Tables quick facts from 2x to 12x is more likely to:

- get ahead and gain confidence / speed in maths generally from now on

- gain easier access to, and understanding of future work on fractions, decimals, percentages, ratio, equations, algebra, time-distance-speed problems, perimeter, area, volume, mass... and of course multiplication / division problems.

As a teacher, the question I have been asked most often is, "How can I help my child learn their Times Tables?"

Writing them out in lists, chanting them, having them on wall posters... these methods are all known to us, but often just don't seem to be enough.

Here is a step by step, systematic approach I have used with students who went on to be able to say, **"I KNOW all my Times Tables quick facts to 12x!"**

It also includes my personal advice on activities I have found to work particularly well for specific Times Tables, and offers ideas to cater for **students' different learning styles**.

DO READ all the information on getting started, record keeping, guidance for use, ideas and tips provided before you begin the Times Tables pages.

Don't skip it... it's important.

The 'I CAN / I KNOW' Checklist for
Unit 4: Times Tables, Exponents, and Prime Numbers

Page	I CAN / I KNOW	Tick & Date Visited	Achieved
210	I KNOW my 2x Table quick facts		❑
212	I KNOW my 3x Table quick facts		❑
214	I KNOW my 4x Table quick facts		❑
216	I KNOW my 5x Table quick facts		❑
218	I KNOW my 6x Table quick facts		❑
220	I KNOW my 7x Table quick facts		❑
222	I KNOW my 8x Table quick facts		❑
224	I KNOW my 9x Table quick facts		❑
226	I KNOW my 10x Table quick facts		❑
228	I KNOW my 11x Table quick facts		❑
230	I KNOW my 12x Table quick facts		❑
233	I KNOW all my square numbers up to 144 (up to 12 x 12 = 144)		❑
235	I KNOW that square root means a root (start number) multiplied by itself to produce a square number		❑
237	I KNOW that a little ^{2} placed after a number means the number is 'squared'		❑
239	I KNOW that a little ^{3} placed after a number means the number is 'cubed'		❑
(!) 241	I KNOW what prime numbers are		❑
243	I KNOW every Times Table quick fact up to the 12x Table... The Ultimate Times Table Challenge!		❑

Getting Started

For easy record keeping and to track progress, the following pages are particularly central to using this Unit effectively:

- The 'I CAN / I KNOW' checklist (page 196)
- The 'Every Visit Record Chart' (page 198)
- The 'Build Me Up' Times Tables grid (Tip 1: page 201)
- The complete list of 2x to 12x Tables facts (pages 243/244)

Things to consider

The Times Tables programme should be used **regularly** in short bursts.

Start with 2x, 3x, 5x, 10x and 11x Tables, as these tend to be easiest to learn, thus building confidence, then move on to others.

Work gradually and steadily through all the Times Tables and let the student decide which one to target next.

Concerns about progress

Don't worry if the student seems to be **stuck at a Level** (see A, B, C, D Model: page 199) on a Times Table for a few sessions. Just take things slowly and ensure the student is secure at a Level before moving on to the next.

Don't get frustrated if Times Tables learning **is not fast and furious.** If the student needs to count up in steps and is not ready to internalise quick facts, then be happy that s/he can at least get there by counting up; a student might stay at Level A for many sessions.

If certain quick facts are proving to be **stubborn**, isolate them and put them on flash cards to use as warm up activities.

The 'pop-up-finger-method'

This very simple technique is an excellent physical and mental aid for counting up in Times Tables patterns or solving division questions, e.g. 'how many times will 5 fit into...?'

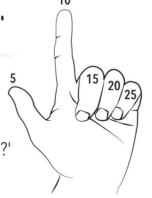
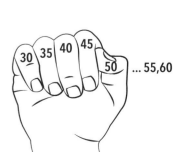

'Every Visit Record Chart'
for Times Tables progress (A, B, C, D Model)

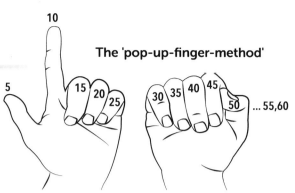

The 'pop-up-finger-method'

	Level A Count up 'out loud' in order, raising a finger / touching nose for each step	Fully Achieved	Level B Count up 'silently' in order, raising a finger for each step, to answer random Times Tables facts	Fully Achieved	Level C Give the answers to random Times Tables facts in, say, less than 10 seconds; if still silently counting up, it is not noticeable	Fully Achieved	Level D Instantly answer random Times Tables facts 'quick fact style' without counting up	Fully Achieved
2x								
3x								
4x								
5x								
6x								
7x								
8x								
9x								
10x								
11x								
12x								

The A, B, C, D Model: guidance for use

What Is It?
It is a systematic programme to learn the Times Tables from 2x to 12x.

There are 4 Levels: A, B, C and D. These have an identical format for all the Times Tables. **Each Times Table has its own copy of the A, B, C, D steps** for easy reference, and is supplemented with suggested activities I find suit the Times Table being studied.

When you first read how to work through the four Levels, it looks laborious, repetitive and over detailed. Stick with it and the process **will** make sense and **does** work.

Students discover a methodical way to LEARN the Times Tables; they can move at a pace which **feels achievable and comfortable**.

I use this systematic method daily; the students like it, and it works. I inject humour and energy into each Level, awarding tokens, stickers and praise for every small step forward.

It has its own 'Every Visit Record Chart' (opposite). It is important to record progress on this chart. My students can tell me which 'Level' they are at for each Times Table; this encourages ownership and responsibility.

How to use it
✓ **Decide** which Times Table is being targeted and turn to the relevant study pages further on in this Unit: always begin at Level A.
✓ Work your way through **Levels A to D** steadily and thoroughly on the chosen Times Table, starting up where you got to in the session before. It may take only one or two sessions to go through Levels A, B, C and D for some Times Tables, and several weeks for others. Work at the learning pace of the student.
✓ Place a dot or tick (dated) on the 'Every Visit Record Chart' (opposite) whenever the student has worked on a Level and use this as a **record of visits and progress**, moving onto the next Level only when the 'Fully Achieved' box is ticked.
✓ Whenever a **Level B** for a Times Table is achieved, enter the answers / products into the **'Build Me Up' Times Tables grid** (Tip 1) and watch it grow.
✓ When more than one Times Table is achieved to Level D, then **combine** them to ask random quick facts, e.g. some 2x, 3x, 5x Tables facts **all mixed in together** in one sitting. Being able to jump from one Times Table to another will improve confidence and begin to build towards the ultimate goal of having 'quick fact recall' for all Times Tables to 12x.
✓ When the student is getting close to completing **all the Times Tables to Level D**, begin to use the **'I KNOW every Times Table quick fact up to the 12x Table'** worksheet (end of this Unit), which lists **all 2x to 12x** Times Tables quick facts.

Ten fabulously simple ideas / tips for you to consider using

The following few pages contain a collection of interesting suggestions, considerations and resource sheets to help make learning Times Tables facts easier and more meaningful.

Students will hopefully find ideas / methods which **appeal to their learning style**.

Select the ideas which appeal the most, and make decisions on how to incorporate them into the programme.

Some of these tips have been around for years and are readily available elsewhere. This selection contains the ones I have found best complement the Bridge the Gap Maths™ programme.

Tip 1   The 'Build Me Up' Times Tables grid
Tip 2   Choosing your preferred Times Table
Tip 3   Bank of Times Tables facts
Tip 4   The Times Tables stick
Tip 5   Families of numbers
Tip 6   The mirror trick
Tip 7   The Times Tables ever reducing steps
Tip 8   Links between Times Tables
Tip 9   Awkward Times Tables facts
Tip 10   Some random ideas to learn Times Tables

Read through the following 10 ideas / tips and discuss each one in detail.

Do this **before** you begin the Times Tables work pages which come afterwards.

 **TIP ① The 'Build Me Up' Times Tables grid**

This is a grid to fill in gradually over the course of learning all the Times Tables from 2x to 12x. Insert the answers / products whenever you achieve Level B status for the Times Table you are working on.

It might take weeks, months... or longer! Be patient, and feel proud as you build them up.

- Fill in the products for the Times Table you have achieved **at Level B**.
- Add others in to the grid as you progress through all the Times Tables.
- Watch it build and grow, and look out for patterns at every opportunity.

x	1	2	3	4	5	6	7	8	9	10	11	12
1												
2												
3												
4												
5												
6												
7												
8												
9												
10												
11												
12												

TIP ② Choosing your preferred Times Table

It is the ultimate goal to be able to give the answer to all the 2x to 12x Tables facts instantly. But, until this level of skill is achieved, a student has to find a 'way' to reach the answer.

Take 9 x 3 as an example:

The student has 2 choices: to use 3x Table or 9x Table knowledge as the starting point. Both will produce the answer, but a student can choose their **preferred** Times Table – the one they find easier and are more confident with, i.e. they may be happier to count up in threes, or feel that the 9x Table 'fingers trick' (see later in this Unit) will get them there faster.

Pointing this out regularly, e.g. by saying, "shall we use the threes or the nines for this?", will remind the student that there are **two routes** to produce the answer.

More examples:

For 7 x 9, choose whether you count up in sevens or use the 9x Table fingers trick.
For 5 x 8, choose whether to count up in fives or eights.
For 12 x 7, choose whether to use the 12x Table Coloured Box Model or count up in sevens.

TIP ③ Bank of Times Tables facts

You need:
* **some thin card (cut up into small pieces)**
* **a sealable food bag**

As you work gradually through the Times Tables, make tiny flash cards of quick facts and keep them in a sealable food bag. Add more and more as you progress. Put 'awkward Times Tables facts' in too, as they will start to sink in sooner. These can be used as a warm up to any maths session, and it is fun to see how many can be placed in a 'Yes I know it' pile or pot.

TIP ④   The Times Tables stick

This is a very simple but highly valuable piece of mathematical equipment, and is really worth making. It is used to point at the Times Tables steps going up in order, backwards or in random order, and is an extremely helpful visual aid to learning Times Tables.

It is excellent for starting Level A for all Times Tables, and is also a good warm up activity, even if not working on Unit 4 at the time.

To make, you need a stick and 2 different colours of tape:

Find a garden stick or similar 'long stick' and make 13 marks on it with coloured tape for 0x at the start and 12x at the end. My stick is 120 cm long, but it does not have to be exactly this length.

Use a different coloured tape for 10x and 5x (because 5x is half of the easy 10x); this will make them stand out visually. Ten times is easy and comforting, and five times is just half of that.

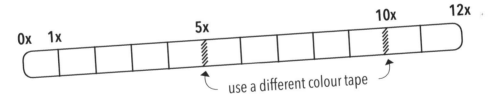

use a different colour tape

To use:
1. Stand opposite the student, holding the stick (horizontally) with one hand in the middle.
2. Decide which Times Table you are working on.
3. Start at the 0x mark and point to each step, asking the student to say the steps (out loud) as you move along the stick (e.g. 7x Table would be 0, 7, 14, 21, 28 etc.) all the way until you get to the other end which is 12x.
4. Do in order from zero to 12x over and over.

When confidence is growing:
- Try going backwards starting at 12x.
- Be mischievous and point to random steps, jumping around and waiting for the student to say the product out loud. Discuss the value of using the 10x and 5x marks as easy visual starting points.
- Hold the stick vertically sometimes, working up and down instead.

When you multiply 2 numbers, it does not matter about the order. The answer will be the same, e.g. 5 x 3 = 15 and 3 x 5 = 15; this means you only have one fact to learn, not two.

Think of this triangle image when you multiply 2 numbers like above: the 3, 5 and 15 are in a family.

3 x 5 = 15
5 x 3 = 15
15 ÷ 5 = 3
15 ÷ 3 = 5

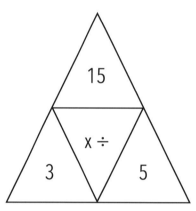

This idea shows how division facts are related, so starting with the answer to a multiplication question (which is called the PRODUCT) and asking the student to give the other 2 numbers (which are called the FACTORS), means that division quick facts are starting to come too, e.g. 15 ÷ 5 = 3 and 15 ÷ 3 = 5 .

This shows that **division** (which often scares students) is directly related to Times Tables. Multiplication and division are the **inverse operations** of each other.

To solve a **division** question, we use **Times Tables** knowledge.

In a Times Table (multiplication) grid, there is a clever MIRROR TRICK happening.

Look at the diagonally highlighted numbers – these are called **square numbers**. The 'triangle' on the left of the grid is a **mirror image** of the triangle on the right. Take a close look.

e.g. 5 x 8 is the same as 8 x 5. They both hit 40. Two facts become one.

x	1	2	3	4	5	6	7	8	9	10	11	12
1	1	2	3	4	5	6	7	8	9	10	11	12
2	2	4	6	8	10	12	14	16	18	20	22	24
3	3	6	9	12	15	18	21	24	27	30	33	36
4	4	8	12	16	20	24	28	32	36	40	44	48
5	5	10	15	20	25	30	35	40	45	50	55	60
6	6	12	18	24	30	36	42	48	54	60	66	72
7	7	14	21	28	35	42	49	56	63	70	77	84
8	8	16	24	32	40	48	56	64	72	80	88	96
9	9	18	27	36	45	54	63	72	81	90	99	108
10	10	20	30	40	50	60	70	80	90	100	110	120
11	11	22	33	44	55	66	77	88	99	110	121	132
12	12	24	36	48	60	72	84	96	108	120	132	144

This means you don't have as many Times Tables facts to learn as you thought!

Do say a Times Table fact *both* ways:
    e.g.   Sometimes say, "5 x 8" = 40
               Sometimes say, "8 x 5" = 40

Also, sometimes start with 40 (product / answer) to supply factors of 5 and 8.
    e.g.   40 = ? x 8   *or*   How many 8s in 40?

Excluding the 1x Table, here are all the Times Tables you will ever need up to 12 x 12.

This works because once you know, say 7 x 4, then of course 4 x 7 has the same answer and is the **same quick fact**.

Remember to **say them backwards** sometimes so you get used to hearing them both ways.

Look at how they make a step pattern as each Times Table has **one less entry** in its column; to figure out why, look first at how we are able to **miss out 2 x 3 in the 3x Table column because 3 x 2 is already there in the 2x Table column**. Fabulous!

2 x 2 = 4				
3 x 2 = 6	3 x 3 = 9			
4 x 2 = 8	4 x 3 = 12	4 x 4 = 16		
5 x 2 = 10	5 x 3 = 15	5 x 4 = 20	5 x 5 = 25	
6 x 2 = 12	6 x 3 = 18	6 x 4 = 24	6 x 5 = 30	6 x 6 = 36
7 x 2 = 14	7 x 3 = 21	7 x 4 = 28	7 x 5 = 35	7 x 6 = 42
8 x 2 = 16	8 x 3 = 24	8 x 4 = 32	8 x 5 = 40	8 x 6 = 48
9 x 2 = 18	9 x 3 = 27	9 x 4 = 36	9 x 5 = 45	9 x 6 = 54
10 x 2 = 20	10 x 3 = 30	10 x 4 = 40	10 x 5 = 50	10 x 6 = 60
11 x 2 = 22	11 x 3 = 33	11 x 4 = 44	11 x 5 = 55	11 x 6 = 66
12 x 2 = 24	12 x 3 = 36	12 x 4 = 48	12 x 5 = 60	12 x 6 = 72

7 x 7 = 49				
8 x 7 = 56	8 x 8 = 64			
9 x 7 = 63	9 x 8 = 72	9 x 9 = 81		
10 x 7 = 70	10 x 8 = 80	10 x 9 = 90	10 x 10 = 100	
11 x 7 = 77	11 x 8 = 88	11 x 9 = 99	11 x 10 = 110	11 x 11 = 121
12 x 7 = 84	12 x 8 = 96	12 x 9 = 108	12 x 10 = 120	12 x 11 = 132	12 x 12 = 144

Links between Times Tables

Remember that there are helpful links between different Times Tables.

For example, if you know your 2x Table, then you already have the first six answers for the 4x Table:

2x:	2	4	6	8	10	12	14	16	18	20	22	24
		↓		↓		↓		↓		↓		↓
4x:		4		8		12		16		20		24

(this takes you up to 6 x 4)

3x:	3	6	9	12	15	18	21	24	27	30	33	36
		↓		↓		↓		↓		↓		↓
6x:		6		12		18		24		30		36

(this takes you up to 6 x 6)

5x:	5	10	15	20	25	30	35	40	45	50	55	60
		↓		↓		↓		↓		↓		↓
10x:		10		20		30		40		50		60

(this takes you up to 6 x 10)

6x:	6	12	18	24	30	36	42	48	54	60	66	72
		↓		↓		↓		↓		↓		↓
12x:		12		24		36		48		60		72

(this takes you up to 6 x 12)

I call them 'the hardest Times Tables facts in the world'.

Despite learning Times Tables carefully and making good progress, there are a few facts which just seem to be the **awkward ones**.

Lots of students seem to find these particular ones hard to remember.

In my experience, these are:

11 x 11 = 121

11 x 12 = 132
and
12 x 11 = 132

9 x 12 = 108
and
12 x 9 = 108

7 x 8 = 56
and
8 x 7 = 56

6 x 7 = 42
and
7 x 6 = 42

6 x 8 = 48
and
8 x 6 = 48

So, give yourself a head start!

- Write these tricky Times Tables facts out on pieces of card and stick them on the fridge door **now**, (and anywhere else where you will see them regularly).
- Say them / test them daily until you have them in your brain.
- If you find other ones are causing you problems, add them to the list.

TIP ⑩ Some random ideas to learn Times Tables

And finally, maybe these other ideas for learning your Times Tables will really appeal to you.

Select your Times Table target and...

- move a counter up a 100 square games board in the Times Table steps
- sing it to a simple or catchy well known tune
- compose and perform a Times Tables rap
- play 'Throw and Catch Q and A Times Tables' with a ball
- design and make 'match the answer' card games
- set a 'learn by this date challenge' with rewards
- write it out on a large poster for the wall
- have 'writing it out on paper' speed challenges
- have 'saying it out loud' speed challenges
- play Times Tables games online
- see how many you can answer in 3 minutes
- reverse student / helper status: student asks, helper answers
- look for Times Tables prompt / flash cards in stores (or make your own)
- design and play a Times Tables board game
- make colourful flash cards for the fridge door
- write Times Tables out in neat lists, adding colours for answers
- play 'Times Tables Challenge' in the car or on walks
- sit on the floor and play 'Roll the Tennis Ball' Q and A
- regularly refer to and use Times Tables grids
- make a Times Table snake with long thin paper strips
- make a Times Table facts number spinner
- add to this list: be creative

Discover the best way to suit your learning style!

I KNOW my 2x Table quick facts

2x Table: Levels A, B, C and D

Level A: counting up in twos using fingers as visual aid.

- Student holds out both hands in front at eye level and spreads 10 fingers, thumbs on outside. Curl fingers into fists. Pop up first thumb, saying, "two", second finger for "four" etc. until get to 24 (nodding for 11x and 12x). Move to helper saying, "1 times 2 is..., 2 times 2 is..." where student pops a finger up and answers. Repeat.

    FOR VARIETY   Hold hands up to the face to touch nose with each finger OR
    helper touch fingers in order (can be fun / visually helpful to try reverse order).

- Helper points to a random finger; student counts up out loud, in order, until reaches the answer.
- When the student can **count up in order confidently**, tick off Level A and move to B.

Level B: beginning to wean off 'out loud and visual' counting up.

- Place the hands in same position as in Level A but now on thighs.
- Helper asks any 2x Table fact; student counts up in order silently, **not looking down** at the fingers as they pop up, and gives the answer. This helps to 'feel' the steps.
- When the student can **give the answer to a random 2x Table question** after silent counting up / popping up fingers, tick off Level B and move to C.

> Flip back to Tip ① once you've achieved this level!

Have you filled the 2x Table lines in the 'Build Me Up' Times Tables grid? Watch it grow........

Level C: to internalise the process and move towards 'just knowing' the answer.

- Student sits on hands. Helper asks any 2x Table fact; student has a few silent seconds to mentally find the answer. If still mentally counting up, it is not noticeable.
- When the student can **answer in a few seconds**, tick off Level C and move to D.

Level D: to instantly answer any 2x Table fact without counting up.

- Helper asks any 2x Table fact and student answers immediately.
- When the student can **instantly answer any 2x Table fact**, tick off Level D and decide which Times Table is next.
- Remember to orally combine with other Times Tables achieved so far.

210

Table: Tips and Ideas

* 2x means **'doubled'**, or the number **added to itself**, e.g. 2 x 7 is 7 + 7
* 2x Table answers always **end in 0, 2, 4, 6 or 8**
* Use the Times Tables stick to count up, back and randomly in twos
* **'2, 4, 6, 8, Who do we appreciate? 10, 12, 14, 16, counting twos is really great! 18, 20, 22, 24 now we know it, close the door!'**
* Play the **'How fast can I write** the 2x Table facts in a list?' game
* Count up in a **rhythmic chant** 2, 4, 6, 8... to 24 / Count back 24, 22, 20, 18... to zero
* Say out loud: '1 times 2 is 2, 2 times 2 is 4, 3 times 2 is 6... to 12 x 2 is 24'
* Point out on the Times Tables stick that **10 x 2 is 20**, so 5 x 2 is **simply half of that**: 10

1 x 2 = 2	12 x 2 = 24	9 x 2 = 18
2 x 2 = 4	11 x 2 = 22	4 x 2 = 8
3 x 2 = 6	10 x 2 = 20	10 x 2 = 20
4 x 2 = 8	9 x 2 = 18	7 x 2 = 14
5 x 2 = 10	8 x 2 = 16	1 x 2 = 2
6 x 2 = 12	7 x 2 = 14	6 x 2 = 12
7 x 2 = 14	6 x 2 = 12	8 x 2 = 16
8 x 2 = 16	5 x 2 = 10	3 x 2 = 6
9 x 2 = 18	4 x 2 = 8	11 x 2 = 22
10 x 2 = 20	3 x 2 = 6	2 x 2 = 4
11 x 2 = 22	2 x 2 = 4	12 x 2 = 24
12 x 2 = 24	1 x 2 = 2	5 x 2 = 10

Try

Fill in the 2x Table strips, colour them in, and say them out loud.

x	1	2	3	4	5	6	7	8	9	10	11	12
1	1		3	4	5	6	7	8	9	10	11	12
2												
3	3		9	12	15	18	21	24	27	30	33	36
4	4		12	16	20	24	28	32	36	40	44	48
5	5		15	20	25	30	35	40	45	50	55	60
6	6		18	24	30	36	42	48	54	60	66	72
7	7		21	28	35	42	49	56	63	70	77	84
8	8		24	32	40	48	56	64	72	80	88	96
9	9		27	36	45	54	63	72	81	90	99	108
10	10		30	40	50	60	70	80	90	100	110	120
11	11		33	44	55	66	77	88	99	110	121	132
12	12		36	48	60	72	84	96	108	120	132	144

3x Table: Levels A, B, C and D *(... but I always start with the rap idea opposite)*

Level A: counting up in threes using fingers as visual aid.

- Student holds out both hands in front at eye level and spreads 10 fingers, thumbs on outside. Curl fingers into fists. Pop up first thumb, saying, "three", second finger for "six" etc. until get to 36 (nodding for 11x and 12x). Move to helper saying, "1 times 3 is…, 2 times 3 is…" where student pops a finger up and answers. Repeat.

  FOR VARIETY   Hold hands up to the face to touch nose with each finger OR
  helper touch fingers in order (can be fun / visually helpful to try reverse order).

- Helper points to a random finger; student counts up out loud, in order, until reaches the answer.
- When the student can **count up in order confidently**, tick off Level A and move to B.

Level B: beginning to wean off 'out loud and visual' counting up.

- Place the hands in same position as in Level A but now on thighs.
- Helper asks any 3x Table fact; student counts up in order silently, **not looking down** at the fingers as they pop up, and gives the answer. This helps to 'feel' the steps.
- When the student can **give the answer to a random 3x Table question** after silent counting up / popping up fingers, tick off Level B and move to C.

> Flip back to Tip ① once you've achieved this level!

Have you filled the 3x Table lines in the 'Build Me Up' Times Tables grid? Watch it grow........

Level C: to internalise the process and move towards 'just knowing' the answer.

- Student sits on hands. Helper asks any 3x Table fact; student has a few silent seconds to mentally find the answer. If still mentally counting up, it is not noticeable.
- When the student can **answer in a few seconds**, tick off Level C and move to D.

Level D: to instantly answer any 3x Table fact without counting up.

- Helper asks any 3x Table fact and student answers immediately.
- When the student can **instantly answer any 3x Table fact**, tick off Level D and decide which Times Table is next.
- Remember to orally combine with other Times Tables achieved so far.

3x Table: Tips and Ideas

* Compose a 3x Table **rhythmic rap**, with claps / wooden spoon on pans: "3, 6, 9, 12..."
* Make a **large number line** for 3x Table on long paper (old wallpaper) using a ruler and regular intervals: move a button along to say steps or cover then reveal numbers
* Use the Times Tables stick to count up, back, and randomly in threes
* Count up in a rhythmic chant 3, 6, 9, 12... to 36 / Count back 36, 33, 30, 27... to zero
* Play the **'How fast can I write** the 3x Table facts in a list?' game
* Say out loud: '1 times 3 is 3, 2 times 3 is 6, 3 times 3 is 9... to 12 x 3 is 36'
* Point out on Times Tables stick that **10 x 3 is 30**, so 5 x 3 is **simply half of that**: 15

1 x 3 = 3	12 x 3 = 36	9 x 3 = 27
2 x 3 = 6	11 x 3 = 33	4 x 3 = 12
3 x 3 = 9	10 x 3 = 30	10 x 3 = 30
4 x 3 = 12	9 x 3 = 27	7 x 3 = 21
5 x 3 = 15	8 x 3 = 24	1 x 3 = 3
6 x 3 = 18	7 x 3 = 21	6 x 3 = 18
7 x 3 = 21	6 x 3 = 18	8 x 3 = 24
8 x 3 = 24	5 x 3 = 15	3 x 3 = 9
9 x 3 = 27	4 x 3 = 12	11 x 3 = 33
10 x 3 = 30	3 x 3 = 9	2 x 3 = 6
11 x 3 = 33	2 x 3 = 6	12 x 3 = 36
12 x 3 = 36	1 x 3 = 3	5 x 3 = 15

Try

Fill in the 3x Table strips, colour them in, and say them out loud.

x	1	2	3	4	5	6	7	8	9	10	11	12
1	1	2		4	5	6	7	8	9	10	11	12
2	2	4		8	10	12	14	16	18	20	22	24
3												
4	4	8		16	20	24	28	32	36	40	44	48
5	5	10		20	25	30	35	40	45	50	55	60
6	6	12		24	30	36	42	48	54	60	66	72
7	7	14		28	35	42	49	56	63	70	77	84
8	8	16		32	40	48	56	64	72	80	88	96
9	9	18		36	45	54	63	72	81	90	99	108
10	10	20		40	50	60	70	80	90	100	110	120
11	11	22		44	55	66	77	88	99	110	121	132
12	12	24		48	60	72	84	96	108	120	132	144

4x Table: Levels A, B, C and D (... but I always start with the song idea opposite)

Level A: counting up in fours using fingers as visual aid.

- Student holds out both hands in front at eye level and spreads 10 fingers, thumbs on outside. Curl fingers into fists. Pop up first thumb, saying, "four", second finger for "eight" etc. until get to 48 (nodding for 11x and 12x). Move to helper saying, "1 times 4 is..., 2 times 4 is..." where student pops a finger up and answers. Repeat.

  FOR VARIETY   Hold hands up to the face to touch nose with each finger OR
  helper touch fingers in order (can be fun / visually helpful to try reverse order).

- Helper points to a random finger; student counts up out loud, in order, until reaches the answer.
- When the student can **count up in order confidently**, tick off Level A and move to B.

Level B: beginning to wean off 'out loud and visual' counting up.

- Place the hands in same position as in Level A but now on thighs.
- Helper asks any 4x Table fact; student counts up in order silently, **not looking down** at the fingers as they pop up, and gives the answer. This helps to 'feel' the steps.
- When the student can **give the answer to a random 4x Table question** after silent counting up / popping up fingers, tick off Level B and move to C.

> Flip back to Tip ① once you've achieved this level!

Have you filled the 4x Table lines in the 'Build Me Up' Times Tables grid? Watch it grow......

Level C: to internalise the process and move towards 'just knowing' the answer.

- Student sits on hands. Helper asks any 4x Table fact; student has a few silent seconds to mentally find the answer. If still mentally counting up, it is not noticeable.
- When the student can **answer in a few seconds**, tick off Level C and move to D.

Level D: to instantly answer any 4x Table fact without counting up.

- Helper asks any 4x Table fact and student answers immediately.
- When the student can **instantly answer any 4x Table fact**, tick off Level D and decide which Times Table is next.
- Remember to orally combine with other Times Tables achieved so far.

4x Table: Tips and Ideas

* **4x Table Song**: sing to 'Twinkle Twinkle Little Star, How I Wonder What You Are!' (4, 8, 12, 16, 20, 24, 28, 32, 36, 40 but 'speak' 44 and 48)
* Say the **2x Table** but shout for the **4x Table 'matches'** (as every other one in the 2x Table is in the 4x Table: 2, **4**, 6, **8**, 10, **12**, 14, **16** etc.)
* **Compare 4x and 2x grids**: notice they will **share** some answers and discuss why
* Use the Times Tables stick to count up, back and randomly in fours
* Play the '**How fast can I write** the 4x Table facts in a list?' game
* Say out loud: '1 times 4 is 4, 2 times 4 is 8, 3 times 4 is 12... to 12 x 4 is 48'
* Play the **sit / stand** game: **sit** '4', **stand** '8', **sit** '12', **stand** '16'...
* Point out on the Times Tables stick that **10 x 4 is 40**, so 5 x 4 is **simply half of that**: 20

1 x 4 = 4	12 x 4 = 48	9 x 4 = 36
2 x 4 = 8	11 x 4 = 44	4 x 4 = 16
3 x 4 = 12	10 x 4 = 40	10 x 4 = 40
4 x 4 = 16	9 x 4 = 36	7 x 4 = 28
5 x 4 = 20	8 x 4 = 32	1 x 4 = 4
6 x 4 = 24	7 x 4 = 28	6 x 4 = 24
7 x 4 = 28	6 x 4 = 24	8 x 4 = 32
8 x 4 = 32	5 x 4 = 20	3 x 4 = 12
9 x 4 = 36	4 x 4 = 16	11 x 4 = 44
10 x 4 = 40	3 x 4 = 12	2 x 4 = 8
11 x 4 = 44	2 x 4 = 8	12 x 4 = 48
12 x 4 = 48	1 x 4 = 4	5 x 4 = 20

Try

Fill in the 4x Table strips, colour them in, and say them out loud.

x	1	2	3	4	5	6	7	8	9	10	11	12
1	1	2	3		5	6	7	8	9	10	11	12
2	2	4	6		10	12	14	16	18	20	22	24
3	3	6	9		15	18	21	24	27	30	33	36
4												
5	5	10	15		25	30	35	40	45	50	55	60
6	6	12	18		30	36	42	48	54	60	66	72
7	7	14	21		35	42	49	56	63	70	77	84
8	8	16	24		40	48	56	64	72	80	88	96
9	9	18	27		45	54	63	72	81	90	99	108
10	10	20	30		50	60	70	80	90	100	110	120
11	11	22	33		55	66	77	88	99	110	121	132
12	12	24	36		60	72	84	96	108	120	132	144

I KNOW my 5x Table quick facts

5x Table: Levels A, B, C and D *(... I find Level A is often in place already)*

Level A: counting up in fives using fingers as visual aid.

- Student holds out both hands in front at eye level and spreads 10 fingers, thumbs on outside. Curl fingers into fists. Pop up first thumb, saying, "five", second finger for "ten" etc. until get to 60 (nodding for 11x and 12x). Move to helper saying, "1 times 5 is..., 2 times 5 is..." where student pops a finger up and answers. Repeat.

    FOR VARIETY   Hold hands up to the face to touch nose with each finger OR
    helper touch fingers in order (can be fun / visually helpful to try reverse order).

- Helper points to a random finger; student counts up out loud, in order, until reaches the answer.
- When the student can **count up in order confidently**, tick off Level A and move to B.

Level B: beginning to wean off 'out loud and visual' counting up.

- Place the hands in same position as in Level A but now on thighs.
- Helper asks any 5x Table fact; student counts up in order silently, **not looking down** at the fingers as they pop up, and gives the answer. This helps to 'feel' the steps.
- When the student can **give the answer to a random 5x Table question** after silent counting up / popping up fingers, tick off Level B and move to C.

> Flip back to Tip ① once you've achieved this level!

Have you filled the 5x Table lines in the 'Build Me Up' Times Tables grid? Watch it grow.........

Level C: to internalise the process and move towards 'just knowing' the answer.

- Student sits on hands. Helper asks any 5x Table fact; student has a few silent seconds to mentally find the answer. If still mentally counting up, it is not noticeable.
- When the student can **answer in a few seconds**, tick off Level C and move to D.

Level D: to instantly answer any 5x Table fact without counting up.

- Helper asks any 5x Table fact and student answers immediately.
- When the student can **instantly answer any 5x Table fact**, tick off Level D and decide which Times Table is next.
- Remember to orally combine with other Times Tables achieved so far.

216

5x Table: Tips and Ideas

* 5x Table answers always end in **0 or 5**
* As most find this Times Table easy, play **'How Fast and Far Can You Go?'**
* If you are **stuck on ANY 5x** Table question, e.g. 5 x 42, **find 10x** and then simply **half it**
* Hold up a **high five** with alternate hands for each step
* Link the 5x Table to the **minute hand** on a clock face
* Use the Times Tables stick to count up, back and randomly in fives
* Play the '**How fast can I write** the 5x Table facts in a list?' game

1 x 5 = 5	12 x 5 = 60	9 x 5 = 45
2 x 5 = 10	11 x 5 = 55	4 x 5 = 20
3 x 5 = 15	10 x 5 = 50	10 x 5 = 50
4 x 5 = 20	9 x 5 = 45	7 x 5 = 35
5 x 5 = 25	8 x 5 = 40	1 x 5 = 5
6 x 5 = 30	7 x 5 = 35	6 x 5 = 30
7 x 5 = 35	6 x 5 = 30	8 x 5 = 40
8 x 5 = 40	5 x 5 = 25	3 x 5 = 15
9 x 5 = 45	4 x 5 = 20	11 x 5 = 55
10 x 5 = 50	3 x 5 = 15	2 x 5 = 10
11 x 5 = 55	2 x 5 = 10	12 x 5 = 60
12 x 5 = 60	1 x 5 = 5	5 x 5 = 25

Try

Fill in the 5x Table strips, colour them in, and say them out loud.

x	1	2	3	4	5	6	7	8	9	10	11	12
1	1	2	3	4		6	7	8	9	10	11	12
2	2	4	6	8		12	14	16	18	20	22	24
3	3	6	9	12		18	21	24	27	30	33	36
4	4	8	12	16		24	28	32	36	40	44	48
5												
6	6	12	18	24		36	42	48	54	60	66	72
7	7	14	21	28		42	49	56	63	70	77	84
8	8	16	24	32		48	56	64	72	80	88	96
9	9	18	27	36		54	63	72	81	90	99	108
10	10	20	30	40		60	70	80	90	100	110	120
11	11	22	33	44		66	77	88	99	110	121	132
12	12	24	36	48		72	84	96	108	120	132	144

6x Table: Levels A, B, C and D

Level A: counting up in sixes using fingers as visual aid.

- Student holds out both hands in front at eye level and spreads 10 fingers, thumbs on outside. Curl fingers into fists. Pop up first thumb, saying, "six", second finger for "twelve" etc. until get to 72 (nodding for 11x and 12x). Move to helper saying, "1 times 6 is…, 2 times 6 is…" where student pops a finger up and answers. Repeat.

   FOR VARIETY   Hold hands up to the face to touch nose with each finger OR
   helper touch fingers in order (can be fun / visually helpful to try reverse order).

- Helper points to a random finger; student counts up out loud, in order, until reaches the answer.
- When the student can **count up in order confidently**, tick off Level A and move to B.

Level B: beginning to wean off 'out loud and visual' counting up.

- Place the hands in same position as in Level A but now on thighs.
- Helper asks any 6x Table fact; student counts up in order silently, **not looking down** at the fingers as they pop up, and gives the answer. This helps to 'feel' the steps.
- When the student can **give the answer to a random 6x Table question** after silent counting up / popping up fingers, tick off Level B and move to C.

> Flip back to Tip ① once you've achieved this level!

Have you filled the 6x Table lines in the 'Build Me Up' Times Tables grid? Watch it grow…….

Level C: to internalise the process and move towards 'just knowing' the answer.

- Student sits on hands. Helper asks any 6x Table fact; student has a few silent seconds to mentally find the answer. If still mentally counting up, it is not noticeable.
- When the student can **answer in a few seconds**, tick off Level C and move to D.

Level D: to instantly answer any 6x Table fact without counting up.

- Helper asks any 6x Table fact and student answers immediately.
- When the student can **instantly answer any 6x Table fact**, tick off Level D and decide which Times Table is next.
- Remember to orally combine with other Times Tables achieved so far.

6x Table: Tips and Ideas

* Make a 6x Table **rap or song** using a strong beat
* Say the **3x Table** but **shout** for the **6x Table 'matches'**
  (as every other one in the 3x Table is in the 6x Table: 3, **6**, 9, **12**, 15, **18**, 21, **24** etc.)
* **Compare 6x and 3x grids**: notice they will **share** some answers and discuss why
* Use the Times Tables stick to count up, back and randomly in sixes
* Count up in a **rhythmic chant** 6, 12, 18, 24... to 72 / Count back 72, 66, 54, 48... to zero
* Play the '**How fast can I write** the 6x Table facts in a list?' game
* Say out loud: '1 times 6 is 6, 2 times 6 is 12, 3 times 6 is 18... to 12 x 6 is 72'
* Point out on the Times Tables stick that **10 x 6 is 60**, so 5 x 6 is **simply half of that**: 30

1 x 6 = 6	12 x 6 = 72	9 x 6 = 54
2 x 6 = 12	11 x 6 = 66	4 x 6 = 24
3 x 6 = 18	10 x 6 = 60	10 x 6 = 60
4 x 6 = 24	9 x 6 = 54	7 x 6 = 42
5 x 6 = 30	8 x 6 = 48	1 x 6 = 6
6 x 6 = 36	7 x 6 = 42	6 x 6 = 36
7 x 6 = 42	6 x 6 = 36	8 x 6 = 48
8 x 6 = 48	5 x 6 = 30	3 x 6 = 18
9 x 6 = 54	4 x 6 = 24	11 x 6 = 66
10 x 6 = 60	3 x 6 = 18	2 x 6 = 12
11 x 6 = 66	2 x 6 = 12	12 x 6 = 72
12 x 6 = 72	1 x 6 = 6	5 x 6 = 30

Try

Fill in the 6x Table strips, colour them in, and say them out loud.

x	1	2	3	4	5	6	7	8	9	10	11	12
1	1	2	3	4	5		7	8	9	10	11	12
2	2	4	6	8	10		14	16	18	20	22	24
3	3	6	9	12	15		21	24	27	30	33	36
4	4	8	12	16	20		28	32	36	40	44	48
5	5	10	15	20	25		35	40	45	50	55	60
6												
7	7	14	21	28	35		49	56	63	70	77	84
8	8	16	24	32	40		56	64	72	80	88	96
9	9	18	27	36	45		63	72	81	90	99	108
10	10	20	30	40	50		70	80	90	100	110	120
11	11	22	33	44	55		77	88	99	110	121	132
12	12	24	36	48	60		84	96	108	120	132	144

I KNOW my 7x Table quick facts

7x Table: Levels A, B, C and D *(... but I always start with the washing line idea opposite)*

Level A: counting up in sevens using fingers as visual aid.

- Student holds out both hands in front at eye level and spreads 10 fingers, thumbs on outside. Curl fingers into fists. Pop up first thumb, saying, "seven", second finger for "fourteen" etc. until get to 84 (nodding for 11x and 12x). Move to helper saying, "1 times 7 is..., 2 times 7 is..." where student pops a finger up and answers. Repeat.

  FOR VARIETY   Hold hands up to the face to touch nose with each finger OR
  helper touch fingers in order (can be fun / visually helpful to try reverse order).

- Helper points to a random finger; student counts up out loud, in order, until reaches the answer.
- When the student can **count up in order confidently**, tick off Level A and move to B.

Level B: beginning to wean off 'out loud and visual' counting up.

- Place the hands in same position as in Level A but now on thighs.
- Helper asks any 7x Table fact; student counts up in order silently, **not looking down** at the fingers as they pop up, and gives the answer. This helps to 'feel' the steps.
- When the student can **give the answer to a random 7x Table question** after silent counting up / popping up fingers, tick off Level B and move to C.

> Flip back to Tip ① once you've achieved this level!

Have you filled the 7x Table lines in the 'Build Me Up' Times Tables grid? Watch it grow........

Level C: to internalise the process and move towards 'just knowing' the answer.

- Student sits on hands. Helper asks any 7x Table fact; student has a few silent seconds to mentally find the answer. If still mentally counting up, it is not noticeable.
- When the student can **answer in a few seconds**, tick off Level C and move to D.

Level D: to instantly answer any 7x Table fact without counting up.

- Helper asks any 7x Table fact and student answers immediately.
- When the student can **instantly answer any 7x Table fact**, tick off Level D and decide which Times Table is next.
- Remember to orally combine with other Times Tables achieved so far.

7x Table: Tips and Ideas

* Make a **washing line with pegs** and 7x Table **answer flags** like party bunting
* Use the Times Tables stick to count up, back and randomly in sevens
* Count up in a **rhythmic chant** 7, 14, 21, 28… to 84 / Count back 84, 77, 70, 63… to zero
* Play the '**How fast can I write** the 7x Table facts in a list?' game
* Say out loud: '1 times 7 is 7, 2 times 7 is 14, 3 times 7 is 21… to 12 x 7 is 84'
* Point out on Times Tables stick that **10 x 7 is 70**, so 5 x 7 is **simply half of that**: 35

1 x 7 = 7	12 x 7 = 84	9 x 7 = 63
2 x 7 = 14	11 x 7 = 77	4 x 7 = 28
3 x 7 = 21	10 x 7 = 70	10 x 7 = 70
4 x 7 = 28	9 x 7 = 63	7 x 7 = 49
5 x 7 = 35	8 x 7 = 56	1 x 7 = 7
6 x 7 = 42	7 x 7 = 49	6 x 7 = 42
7 x 7 = 49	6 x 7 = 42	8 x 7 = 56
8 x 7 = 56	5 x 7 = 35	3 x 7 = 21
9 x 7 = 63	4 x 7 = 28	11 x 7 = 77
10 x 7 = 70	3 x 7 = 21	2 x 7 = 14
11 x 7 = 77	2 x 7 = 14	12 x 7 = 84
12 x 7 = 84	1 x 7 = 7	5 x 7 = 35

Try

Fill in the 7x Table strips, colour them in, and say them out loud.

x	1	2	3	4	5	6	7	8	9	10	11	12
1	1	2	3	4	5	6		8	9	10	11	12
2	2	4	6	8	10	12		16	18	20	22	24
3	3	6	9	12	15	18		24	27	30	33	36
4	4	8	12	16	20	24		32	36	40	44	48
5	5	10	15	20	25	30		40	45	50	55	60
6	6	12	18	24	30	36		48	54	60	66	72
7												
8	8	16	24	32	40	48		64	72	80	88	96
9	9	18	27	36	45	54		72	81	90	99	108
10	10	20	30	40	50	60		80	90	100	110	120
11	11	22	33	44	55	66		88	99	110	121	132
12	12	24	36	48	60	72		96	108	120	132	144

I KNOW my 8x Table quick facts

8x Table: Levels A, B, C and D *(... but I always start with the awkward facts cards idea opposite)*

Level A: counting up in eights using fingers as visual aid.

- Student holds out both hands in front at eye level and spreads 10 fingers, thumbs on outside. Curl fingers into fists. Pop up first thumb, saying, "eight", second finger for "sixteen" etc. until get to 96 (nodding for 11x and 12x). Move to helper saying, "1 times 8 is…, 2 times 8 is…" where student pops a finger up and answers. Repeat.

   FOR VARIETY　　　Hold hands up to the face to touch nose with each finger OR
   　　　　　　　　　　helper touch fingers in order (can be fun / visually helpful to try reverse order).

- Helper points to a random finger; student counts up out loud, in order, until reaches the answer.
- When the student can **count up in order confidently**, tick off Level A and move to B.

Level B: beginning to wean off 'out loud and visual' counting up.

- Place the hands in same position as in Level A but now on thighs.
- Helper asks any 8x Table fact; student counts up in order silently, **not looking down** at the fingers as they pop up, and gives the answer. This helps to 'feel' the steps.
- When the student can **give the answer to a random 8x Table question** after silent counting up / popping up fingers, tick off Level B and move to C.

> Flip back to Tip ① once you've achieved this level!

Have you filled the 8x Table lines in the 'Build Me Up' Times Tables grid? Watch it grow……..

Level C: to internalise the process and move towards 'just knowing' the answer.

- Student sits on hands. Helper asks any 8x Table fact; student has a few silent seconds to mentally find the answer. If still mentally counting up, it is not noticeable.
- When the student can **answer in a few seconds**, tick off Level C and move to D.

Level D: to instantly answer any 8x Table fact without counting up.

- Helper asks any 8x Table fact and student answers immediately.
- When the student can **instantly answer any 8x Table fact**, tick off Level D and decide which Times Table is next.
- Remember to orally combine with other Times Tables achieved so far.

8x Table: Tips and Ideas

* Make 2 large fact cards for awkward facts 6 x 8 = 48 and 7 x 8 = 56
* Make up a **story**: e.g. A hungry ant eats 8 fleas, then 16 fleas... and illustrate
* Make an 8x Table **flash card / matching game** - call it '**The Hardest Times Table Game**'
* Notice that the **ones column has a pattern: 8, 6, 4, 2, 0, 8, 6, 4, 2** etc.
* Use the Times Tables stick to count up, back and randomly in eights
* Count up in a **rhythmic chant** 8, 16, 24, 32... to 96 / Count back 96, 88, 80, 72... to zero
* Play the '**How fast can I write** the 8x Table facts in a list?' game
* Say out loud: '1 times 8 is 8, 2 times 8 is 16, 3 times 8 is 24... to 12 x 8 is 96'
* Compare **8x and 4x grids**: notice they will **share** some answers and discuss why
* Point out on Times Tables stick that **10 x 8 is 80**, so 5 x 8 is **simply half of that**: 40

6 x 8 = 48
7 x 8 = 56

1 x 8 = 8	12 x 8 = 96	9 x 8 = 72
2 x 8 = 16	11 x 8 = 88	4 x 8 = 32
3 x 8 = 24	10 x 8 = 80	10 x 8 = 80
4 x 8 = 32	9 x 8 = 72	7 x 8 = 56
5 x 8 = 40	8 x 8 = 64	1 x 8 = 8
6 x 8 = 48	7 x 8 = 56	6 x 8 = 48
7 x 8 = 56	6 x 8 = 48	8 x 8 = 64
8 x 8 = 64	5 x 8 = 40	3 x 8 = 24
9 x 8 = 72	4 x 8 = 32	11 x 8 = 88
10 x 8 = 80	3 x 8 = 24	2 x 8 = 16
11 x 8 = 88	2 x 8 = 16	12 x 8 = 96
12 x 8 = 96	1 x 8 = 8	5 x 8 = 40

Try

Fill in the 8x Table strips, colour them in, and say them out loud.

x	1	2	3	4	5	6	7	8	9	10	11	12
1	1	2	3	4	5	6	7		9	10	11	12
2	2	4	6	8	10	12	14		18	20	22	24
3	3	6	9	12	15	18	21		27	30	33	36
4	4	8	12	16	20	24	28		36	40	44	48
5	5	10	15	20	25	30	35		45	50	55	60
6	6	12	18	24	30	36	42		54	60	66	72
7	7	14	21	28	35	42	49		63	70	77	84
8												
9	9	18	27	36	45	54	63		81	90	99	108
10	10	20	30	40	50	60	70		90	100	110	120
11	11	22	33	44	55	66	77		99	110	121	132
12	12	24	36	48	60	72	84		108	120	132	144

I KNOW my 9x Table quick facts

9x Table: Levels A, B, C and D *(... but I always start with the fingers trick idea opposite)*

Level A: counting up in nines using fingers as visual aid.

- Student holds out both hands in front at eye level and spreads 10 fingers, thumbs on outside. Curl fingers into fists. Pop up first thumb, saying, "nine", second finger for "eighteen" etc. until get to 108 (nodding for 11x and 12x). Move to helper saying, "1 times 9 is..., 2 times 9 is..." where student pops a finger up and answers. Repeat.

  FOR VARIETY   Hold hands up to the face to touch nose with each finger OR
  helper touch fingers in order (can be fun / visually helpful to try reverse order).

- Helper points to a random finger; student counts up out loud, in order, until reaches the answer.
- When the student can **count up in order confidently**, tick off Level A and move to B.

Level B: beginning to wean off 'out loud and visual' counting up.

- Place the hands in same position as in Level A but now on thighs.
- Helper asks any 9x Table fact; student counts up in order silently, **not looking down** at the fingers as they pop up, and gives the answer. This helps to 'feel' the steps.
- When the student can **give the answer to a random 9x Table question** after silent counting up / popping up fingers, tick off Level B and move to C.

> Flip back to
> Tip ① once
> you've achieved
> this level!

Have you filled the 9x Table lines in the 'Build Me Up' Times Tables grid? Watch it grow.........

Level C: to internalise the process and move towards 'just knowing' the answer.

- Student sits on hands. Helper asks any 9x Table fact; student has a few silent seconds to mentally find the answer. If still mentally counting up, it is not noticeable.
- When the student can **answer in a few seconds**, tick off Level C and move to D.

Level D: to instantly answer any 9x Table fact without counting up.

- Helper asks any 9x Table fact and student answers immediately.
- When the student can **instantly answer any 9x Table fact**, tick off Level D and decide which Times Table is next.
- Remember to orally combine with other Times Tables achieved so far.

Table: Tips and Ideas

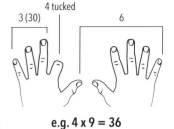

e.g. 4 x 9 = 36

* **9x Fingers Trick**: Lay out hands on a table, palms down, fingers stretched.
  - ☞ **For 2 x 9** tuck in the second finger from the left: the fingers to the left of the tucked finger are tens (**one**), the fingers to the right are ones (**eight**)... **so 18**
  - ☞ **For 3 x 9** tuck in the third finger from the left: the fingers to the left will be **two**, the fingers to the right will be **seven**... **so 27**
  - ☞ It doesn't work for the final 2 facts 11x and 12x; these need to be memorised: 99 and 108
* Notice that the **2 digits in the answers add up to 9**, e.g. 1 + 8 / 2 + 7 / 3 + 6 / 4 + 5 etc. (except 99... but 9 + 9 is 18, and so it then works)
* Use the Times Tables stick to count up, back and randomly in nines
* Play the **'How fast can I write** the 9x Table facts in a list?' game

1 x 9 = 9	12 x 9 = 108	9 x 9 = 81
2 x 9 = 18	11 x 9 = 99	4 x 9 = 36
3 x 9 = 27	10 x 9 = 90	10 x 9 = 90
4 x 9 = 36	9 x 9 = 81	7 x 9 = 63
5 x 9 = 45	8 x 9 = 72	1 x 9 = 9
6 x 9 = 54	7 x 9 = 63	6 x 9 = 54
7 x 9 = 63	6 x 9 = 54	8 x 9 = 72
8 x 9 = 72	5 x 9 = 45	3 x 9 = 27
9 x 9 = 81	4 x 9 = 36	11 x 9 = 99
10 x 9 = 90	3 x 9 = 27	2 x 9 = 18
11 x 9 = 99	2 x 9 = 18	12 x 9 = 108
12 x 9 = 108	1 x 9 = 9	5 x 9 = 45

Try

Fill in the 9x Table strips, colour them in, and say them out loud.

x	1	2	3	4	5	6	7	8	9	10	11	12
1	1	2	3	4	5	6	7	8		10	11	12
2	2	4	6	8	10	12	14	16		20	22	24
3	3	6	9	12	15	18	21	24		30	33	36
4	4	8	12	16	20	24	28	32		40	44	48
5	5	10	15	20	25	30	35	40		50	55	60
6	6	12	18	24	30	36	42	48		60	66	72
7	7	14	21	28	35	42	49	56		70	77	84
8	8	16	24	32	40	48	56	64		80	88	96
9												
10	10	20	30	40	50	60	70	80		100	110	120
11	11	22	33	44	55	66	77	88		110	121	132
12	12	24	36	48	60	72	84	96		120	132	144

10x Table: Levels A, B, C and D *(... often in place, but do check 11 x 10 and 12 x 10)*

Level A: counting up in tens using fingers as visual aid.

- Student holds out both hands in front at eye level and spreads 10 fingers, thumbs on outside. Curl fingers into fists. Pop up first thumb, saying, "ten", second finger for "twenty" etc. until get to 120 (nodding for 11x and 12x). Move to helper saying, "1 times 10 is..., 2 times 10 is..." where student pops a finger up and answers. Repeat.

  FOR VARIETY   Hold hands up to the face to touch nose with each finger OR
  helper touch fingers in order (can be fun / visually helpful to try reverse order).

- Helper points to a random finger; student counts up out loud, in order, until reaches the answer.
- When the student can **count up in order confidently**, tick off Level A and move to B.

Level B: beginning to wean off 'out loud and visual' counting up.

- Place the hands in same position as in Level A but now on thighs.
- Helper asks any 10x Table fact; student counts up in order silently, **not looking down** at the fingers as they pop up, and gives the answer. This helps to 'feel' the steps.
- When the student can **give the answer to a random 10x Table question** after silent counting up / popping up fingers, tick off Level B and move to C.

Flip back to Tip ① once you've achieved this level!

Have you filled the 10x Table lines in the 'Build Me Up' Times Tables grid? Watch it grow......

Level C: to internalise the process and move towards 'just knowing' the answer.

- Student sits on hands. Helper asks any 10x Table fact; student has a few silent seconds to mentally find the answer. If still mentally counting up, it is not noticeable.
- When the student can **answer in a few seconds**, tick off Level C and move to D.

Level D: to instantly answer any 10x Table fact without counting up.

- Helper asks any 10x Table fact and student answers immediately.
- When the student can **instantly answer any 10x Table fact**, tick off Level D and decide which Times Table is next.
- Remember to orally combine with other Times Tables achieved so far.

10x Table: Tips and Ideas

* This Times Table is easy. Simply plop a zero at the end of the start number.
  You can multiply any whole number in the world by ten by just plopping a zero on the end.
  Try it with enormous numbers like 658 or 73,885.
* As most find this Times Table easy, play **'How Fast and Far Can You Go?'**
* Compare **10x and 5x grids**: notice they will **share** some answers and discuss why
* Count up in a **rhythmic chant** 10, 20, 30, 40... to 120 / Count back 120, 110, 100, 90... to zero
* The two 'trickier' 10x facts are **11 x 10 = 110** and **12 x 10 = 120** so put onto cards

| 11 x 10 = 110 |
| 12 x 10 = 120 |

1 x 10 = 10	12 x 10 = 120	9 x 10 = 90
2 x 10 = 20	11 x 10 = 110	4 x 10 = 40
3 x 10 = 30	10 x 10 = 100	10 x 10 = 100
4 x 10 = 40	9 x 10 = 90	7 x 10 = 70
5 x 10 = 50	8 x 10 = 80	1 x 10 = 10
6 x 10 = 60	7 x 10 = 70	6 x 10 = 60
7 x 10 = 70	6 x 10 = 60	8 x 10 = 80
8 x 10 = 80	5 x 10 = 50	3 x 10 = 30
9 x 10 = 90	4 x 10 = 40	11 x 10 = 110
10 x 10 = 100	3 x 10 = 30	2 x 10 = 20
11 x 10 = 110	2 x 10 = 20	12 x 10 = 120
12 x 10 = 120	1 x 10 = 10	5 x 10 = 50

Try

Fill in the 10x Table strips, colour them in, and say them out loud.

x	1	2	3	4	5	6	7	8	9	10	11	12
1	1	2	3	4	5	6	7	8	9		11	12
2	2	4	6	8	10	12	14	16	18		22	24
3	3	6	9	12	15	18	21	24	27		33	36
4	4	8	12	16	20	24	28	32	36		44	48
5	5	10	15	20	25	30	35	40	45		55	60
6	6	12	18	24	30	36	42	48	54		66	72
7	7	14	21	28	35	42	49	56	63		77	84
8	8	16	24	32	40	48	56	64	72		88	96
9	9	18	27	36	45	54	63	72	81		99	108
10												
11	11	22	33	44	55	66	77	88	99		121	132
12	12	24	36	48	60	72	84	96	108		132	144

I KNOW my 11x Table quick facts

11x Table: Levels A, B, C and D *(... but I always start with the number pattern idea opposite)*

Level A: counting up in elevens using fingers as visual aid.

A

- Student holds out both hands in front at eye level and spreads 10 fingers, thumbs on outside. Curl fingers into fists. Pop up first thumb, saying, "eleven", second finger for "twenty two" etc. until get to 132 (nodding for 11x and 12x). Move to helper saying, "1 times 11 is..., 2 times 11 is..." where student pops a finger up and answers. Repeat.

  FOR VARIETY   Hold hands up to the face to touch nose with each finger OR
  helper touch fingers in order (can be fun / visually helpful to try reverse order).

- Helper points to a random finger; student counts up out loud, in order, until reaches the answer.
- When the student can **count up in order confidently**, tick off Level A and move to B.

Level B: beginning to wean off 'out loud and visual' counting up.

B

- Place the hands in same position as in Level A but now on thighs.
- Helper asks any 11x Table fact; student counts up in order silently, **not looking down** at the fingers as they pop up, and gives the answer. This helps to 'feel' the steps.
- When the student can **give the answer to a random 11x Table question** after silent counting up / popping up fingers, tick off Level B and move to C.

> Flip back to Tip ① once you've achieved this level!

Have you filled the 11x Table lines in the 'Build Me Up' Times Tables grid? Watch it grow......

Level C: to internalise the process and move towards 'just knowing' the answer.

C

- Student sits on hands. Helper asks any 11x Table fact; student has a few silent seconds to mentally find the answer. If still mentally counting up, it is not noticeable.
- When the student can **answer in a few seconds**, tick off Level C and move to D.

Level D: to instantly answer any 11x Table fact without counting up.

D

- Helper asks any 11x Table fact and student answers immediately.
- When the student can **instantly answer any 11x Table fact**, tick off Level D and decide which Times Table is next.
- Remember to orally combine with other Times Tables achieved so far.

11x Table: Tips and Ideas

✳ This Times Table is easy up to 9 x 11, as it just repeats the multiplier: 2 x 11 = 22, 6 x 11 = 66 etc.

✳ The three trickier 11x facts are 10x, 11x and 12x. Put them onto cards and discuss these little visual tricks.

| 10 x 11 = 110 | plop a zero on the end of 11 |

| 11 x 11 = 121 | 1-2-1 is a palindrome (same forwards and backwards) |

| 12 x 11 = 132 | The numbers 1, 2 and 3 but out of order! |

✳ Point out on the Times Tables stick that **10 x 11 is 110**, so 5 x 11 is **simply half of that**: 55

1 x 11 = 11	12 x 11 = 132	9 x 11 = 99
2 x 11 = 22	11 x 11 = 121	4 x 11 = 44
3 x 11 = 33	10 x 11 = 110	10 x 11 = 110
4 x 11 = 44	9 x 11 = 99	7 x 11 = 77
5 x 11 = 55	8 x 11 = 88	1 x 11 = 11
6 x 11 = 66	7 x 11 = 77	6 x 11 = 66
7 x 11 = 77	6 x 11 = 66	8 x 11 = 88
8 x 11 = 88	5 x 11 = 55	3 x 11 = 33
9 x 11 = 99	4 x 11 = 44	11 x 11 = 121
10 x 11 = 110	3 x 11 = 33	2 x 11 = 22
11 x 11 = 121	2 x 11 = 22	12 x 11 = 132
12 x 11 = 132	1 x 11 = 11	5 x 11 = 55

Try

Fill in the 11x Table strips, colour them in, and say them out loud.

x	1	2	3	4	5	6	7	8	9	10	11	12
1	1	2	3	4	5	6	7	8	9	10		12
2	2	4	6	8	10	12	14	16	18	20		24
3	3	6	9	12	15	18	21	24	27	30		36
4	4	8	12	16	20	24	28	32	36	40		48
5	5	10	15	20	25	30	35	40	45	50		60
6	6	12	18	24	30	36	42	48	54	60		72
7	7	14	21	28	35	42	49	56	63	70		84
8	8	16	24	32	40	48	56	64	72	80		96
9	9	18	27	36	45	54	63	72	81	90		108
10	10	20	30	40	50	60	70	80	90	100		120
11												
12	12	24	36	48	60	72	84	96	108	120		144

I KNOW my 12x Table quick facts

12x Table: Levels A, B, C and D *(... but I always start with the coloured box model opposite)*

Level A: counting up in twelves using fingers as visual aid.

- Student holds out both hands in front at eye level and spreads 10 fingers, thumbs on outside. Curl fingers into fists. Pop up first thumb, saying, "twelve", second finger for "twenty four" etc. until get to 144 (nodding for 11x and 12x). Move to helper saying, "1 times 12 is..., 2 times 12 is..." where student pops a finger up and answers. Repeat.

  FOR VARIETY   Hold hands up to the face to touch nose with each finger OR
  helper touch fingers in order (can be fun / visually helpful to try reverse order).

- Helper points to a random finger; student counts up out loud, in order, until reaches the answer.
- When the student can **count up in order confidently**, tick off Level A and move to B.

Level B: beginning to wean off 'out loud and visual' counting up.

- Place the hands in same position as in Level A but now on thighs.
- Helper asks any 12x Table fact; student counts up in order silently, **not looking down** at the fingers as they pop up, and gives the answer. This helps to 'feel' the steps.
- When the student can **give the answer to a random 12x Table question** after silent counting up / popping up fingers, tick off Level B and move to C.

> Flip back to
> Tip ① once
> you've achieved
> this level!

Have you filled the 12x Table lines in the 'Build Me Up' Times Tables grid? Watch it grow......

Level C: to internalise the process and move towards 'just knowing' the answer.

- Student sits on hands. Helper asks any 12x Table fact; student has a few silent seconds to mentally find the answer. If still mentally counting up, it is not noticeable.
- When the student can **answer in a few seconds**, tick off Level C and move to D.

Level D: to instantly answer any 12x Table fact without counting up.

- Helper asks any 12x Table fact and student answers immediately.
- When the student can **instantly answer any 12x Table fact**, tick off Level D and decide which Times Table is next.
- Remember to orally combine with other Times Tables achieved so far.

12x Table: Tips and Ideas

* **12x Table Coloured Box Model**: Write out 1 x 12 to 12 x 12 facts under each other in a list.
  * ✎ **Draw a red box** around the first four facts: the answers to these all start with the same start digit as the multiplier (so 1x is **1**2, 2x is **2**4, 3x is **3**6, 4x is **4**8)
  * ✎ **Draw a green box** around the next 5 facts (5x to 9x): the answers all start with a digit **one higher** than the multiplier (so 5x is **six**ty something, 6x is **seven**ty something, 7x is **eigh**ty something, 8x is **nine**ty something and 9x is 'ten**ty' something!)
  * ✎ **Draw a blue box** around the last 3 facts (10x to 12x): as with several Times Tables, you simply have to memorise these (120, 132 & 144).
* Use the Times Tables stick to count up, back and randomly in twelves
* Play the **'How fast can I write** the 12x Table facts in a list?' game
* Point out on the Times Tables stick that **10 x 12 is 120**, so 5 x 12 is **simply half of that**: 60

12 x 12 = 144
The King of
the grid!

1 x 12 = 12	12 x 12 = 144	9 x 12 = 108
2 x 12 = 24	11 x 12 = 132	4 x 12 = 48
3 x 12 = 36	10 x 12 = 120	10 x 12 = 120
4 x 12 = 48	9 x 12 = 108	7 x 12 = 84
5 x 12 = 60	8 x 12 = 96	1 x 12 = 12
6 x 12 = 72	7 x 12 = 84	6 x 12 = 72
7 x 12 = 84	6 x 12 = 72	8 x 12 = 96
8 x 12 = 96	5 x 12 = 60	3 x 12 = 36
9 x 12 = 108	4 x 12 = 48	11 x 12 = 132
10 x 12 = 120	3 x 12 = 36	2 x 12 = 24
11 x 12 = 132	2 x 12 = 24	12 x 12 = 144
12 x 12 = 144	1 x 12 = 12	5 x 12 = 60

Try

Fill in the 12x Table strips, colour them in, and say them out loud.

x	1	2	3	4	5	6	7	8	9	10	11	12
1	1	2	3	4	5	6	7	8	9	10	11	
2	2	4	6	8	10	12	14	16	18	20	22	
3	3	6	9	12	15	18	21	24	27	30	33	
4	4	8	12	16	20	24	28	32	36	40	44	
5	5	10	15	20	25	30	35	40	45	50	55	
6	6	12	18	24	30	36	42	48	54	60	66	
7	7	14	21	28	35	42	49	56	63	70	77	
8	8	16	24	32	40	48	56	64	72	80	88	
9	9	18	27	36	45	54	63	72	81	90	99	
10	10	20	30	40	50	60	70	80	90	100	110	
11	11	22	33	44	55	66	77	88	99	110	121	
12												

The following six 'I CAN / I KNOW' work pages are related directly to Times Tables, Exponents and Prime Numbers.

Give them a go when you feel that your Times Tables confidence is beginning to blossom!

 For tougher skills pages, where I have found in my experience that the student needs to have focussed, ultra high concentration levels, there is a fun 'warning': (!) so you can ensure that the timing is just right to address the few concepts that are maybe a little harder to grasp; in this Unit there is only one for 'prime numbers'.

 For maximum learning impact and ownership of ability:
- Read out loud and discuss the 'I CAN / I KNOW' statement before starting a work page
- Repeat before the 8 'try' questions
- Repeat before the 8 'test' questions
- Finally, repeat once more at the end – then discuss whether you have 'Achieved' or need to revisit later

This helps the student be clear about the learning objective, tune in, focus and take ownership of the learning process. The student him/herself should tick the 'Achieved' box on the record chart.

I KNOW all my square numbers up to 144 (up to 12 x 12 = 144)

A square number is the answer (product) when you multiply a number by itself.

The numbers in bold below are **square numbers**:
Of course, there are many more square numbers – you could go on for ever.

We need to know up to 12 x 12 = **144** as quick facts.

1 x 1 = **1**   7 x 7 = **49**

2 x 2 = **4**   8 x 8 = **64**

3 x 3 = **9**   9 x 9 = **81**

4 x 4 = **16**   10 x 10 = **100**

5 x 5 = **25**   11 x 11 = **121**

6 x 6 = **36**   12 x 12 = **144**

A silly but fun game I use in my 1:1 maths sessions to learn the **square number facts**, is to ask the student to give the answer in a silly voice, or a loud shouting voice. Yes, it is silly, but somehow gives the activity a different status or edge. Younger students especially think it's great fun, and anything that can bring fun to learning Times Tables is valuable!

You could design and make a card game to help you learn these important and useful Times Tables quick facts, or just ask a helper to call them out.

Remember, as well as asking the Times Table question, it is good to **sometimes start with the answer** (the square number itself) and **identify the square ROOT** (the starting number).

Try

Find the square number answers:

What is 6 squared?   7 squared is...?   Multiply 8 by itself   What is eleven squared?

What is 12 x 12?   9 squared is...?   Square the number 5   Is 16 a square number? Why?

Now make up 2 more similar questions of your own!

Test

Find the square number answers:

1.	4 squared is...?	16
2.	9 x 9	81
3.	What is 12 squared?	144
4.	7 x 7	49
5.	3 x 3	9
6.	8 squared is...?	64
7.	6 x 6	36
8.	What is 5 squared?	25

Now re-read the title I KNOW statement; decide whether you have 'Achieved' or need to revisit.

I KNOW that square root means a root (start number) multiplied by itself to produce a square number

 A 'square root' is a number you multiply by itself to produce a 'square number'. If you know your square numbers, then this is just like working in reverse.

'Square root' has its own symbol:

The **square root** of **36** is **6**, because 6 x 6 = 36 $\sqrt{\overset{6}{36}}$

Look at the facts below and place your finger on the square number (the answer / product).

All you have to do is look at **what number was multiplied by itself** to find the **square root**.

1 x 1 = 1	7 x 7 = 49
2 x 2 = 4	8 x 8 = 64
3 x 3 = 9	9 x 9 = 81
4 x 4 = 16	10 x 10 = 100
5 x 5 = 25	11 x 11 = 121
6 x 6 = 36	12 x 12 = 144

TIP
Try out the square root function on your calculator!

Try

Find the square roots of these numbers:

$\sqrt{16}$ $\sqrt{49}$ $\sqrt{144}$ $\sqrt{25}$

$\sqrt{36}$ $\sqrt{121}$ $\sqrt{4}$ $\sqrt{81}$

Now make up 2 more similar questions of your own!

Test

Find the square roots of these numbers:

1. $\sqrt{100}$

10

2. $\sqrt{64}$

8

3. The square root of 81 is...?

9

4. Square root of 16?

4

5. $\sqrt{9}$

3

6. $\sqrt{49}$

7

7. 121 is the square number, so what is the square root?

11

8. $\sqrt{36}$

6

Now re-read the title I KNOW statement; decide whether you have 'Achieved' or need to revisit.

> # I KNOW that a little ² placed after a number means the number is 'squared'

Remember, to square a number, we multiply it by itself. Instead of having to say 'number squared', there is quick way to notate it: we place a little ² after it.

This tiny ² has a special name: it is called an **exponent**. Say this word out loud.

Look at these examples and say them out loud:

2^2 means 2 squared, so 2 x 2 = **4**

3^2 means 3 squared, so 3 x 3 = **9**

4^2 means 4 squared, so 4 x 4 = **16**

5^2 means 5 squared, so 5 x 5 = **25**

6^2 means 6 squared, so 6 x 6 = **36**

7^2 means 7 squared, so 7 x 7 = **49**

8^2 means 8 squared, so 8 x 8 = **64**

9^2 means 9 squared, so 9 x 9 = **81**

10^2 means 10 squared, so 10 x 10 = **100**

11^2 means 11 squared, so 11 x 11 = **121**

12^2 means 12 squared, so 12 x 12 = **144**

Continue the list beyond 12 squared... if you dare!

Try

Find the square numbers:

$5^2 = ?$ $8^2 = ?$ $3^2 = ?$ $7^2 = ?$ $6^2 = ?$ $9^2 = ?$ $4^2 = ?$ $10^2 = ?$

Now make up 2 more similar questions of your own!

Test

Find the square numbers:

1.	$3^2 = ?$	9
2.	$5^2 = ?$	25
3.	$9^2 = ?$	81
4.	$11^2 = ?$	121
5.	$4^2 = ?$	16
6.	$10^2 = ?$	100
7.	$8^2 = ?$	64
8.	$6^2 = ?$	36

Now re-read the title I KNOW statement; decide whether you have 'Achieved' or need to revisit.

I KNOW that a little ^{3} placed after a number means the number is 'cubed'

As well as 'numbers squared' there are also 'numbers cubed'. A number cubed means a number is not just multiplied by itself once, but it is multiplied by itself AGAIN. To notate this, we place a little ^{3} after the number.

This tiny ^{3} has a special name: it is called an **exponent**. Say this word out loud.

Look at these examples and say them out loud:

2^3 means 2 cubed, so 2 x 2 x 2 = **8** (because 2 x 2 is 4, then x 2 again is 8)

3^3 means 3 cubed, so 3 x 3 x 3 = **27** (because 3 x 3 is 9, x 3 again is 27)

4^3 means 4 cubed, so 4 x 4 x 4 = **64** (because 4 x 4 is 16, x 4 again is 64)

5^3 means 5 cubed, so 5 x 5 x 5 = **125** (because 5 x 5 is 25, x 5 again is 125)

6^3 means 6 cubed, so 6 x 6 x 6 = **216** (because 6 x 6 is 36, x 6 again is 216) .... and so on!

On paper, continue the list up to **12 cubed**... or even further if you dare! You may need a calculator.

Notice how finding the square of a number and the cube of a number differ:

5^2 means **5 x 5**: is **25**
5^3 means **5 x 5 x 5**: 5 x 5 is 25, then x 5 again is **125**

Try

Find the cube numbers, perhaps using a calculator:

$3^3 = ?$ $5^3 = ?$ $10^3 = ?$ $2^3 = ?$ $7^3 = ?$ $11^3 = ?$ $8^3 = ?$ $6^3 = ?$

Now make up 2 more similar questions of your own!

Test

Find the cube numbers, perhaps using a calculator:

1.	$3^3 = ?$	$3 \times 3 \times 3 = 27$
2.	$5^3 = ?$	$5 \times 5 \times 5 = 125$
3.	$9^3 = ?$	$9 \times 9 \times 9 = 729$
4.	$11^3 = ?$	$11 \times 11 \times 11 = 1,331$
5.	$4^3 = ?$	$4 \times 4 \times 4 = 64$
6.	$10^3 = ?$	$10 \times 10 \times 10 = 1,000$
7.	$8^3 = ?$	$8 \times 8 \times 8 = 512$
8.	$6^3 = ?$	$6 \times 6 \times 6 = 216$

Now re-read the title I KNOW statement; decide whether you have 'Achieved' or need to revisit.

I KNOW what prime numbers are

A prime number is a special number which is only divisible by itself or 1.

Sounds confusing? Other ways of saying this are:

A prime number...

... will **not allow** any other number (other than itself or 1) to fit into it.

... will **not allow** any other number (other than itself or 1) to be divided into it.

... does **not** have any other divisors (other than itself or 1).

... will **not** appear in any Times Table other than its own and the 1x Table.

e.g. Take the number **11** and put a **circle** around it:

- Will **1** fit in? Yes, of course.
- Will **11** fit in? Yes, of course.
- Use the 'pop-up-finger-method' to count up in 2s, 3s, 4s etc. to see if any other number fits in.
- No! So 11 is prime.

mr. prime number

only me and #1 allowed in my house!

go away other numbers!

Even numbers (except 2 itself) are **not prime** as 2 fits in of course. If a number is **not prime**, we say it is **composite**.

Test these:

not prime

not prime

prime!

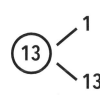
not prime

Try

Identify each of these numbers as 'prime' or 'composite':

12   5   7   15   3   8   11   21

Now make up 2 more similar questions of your own!

Test

Identify each of these numbers as 'prime' or 'composite':

1. 4

composite
(1, 2 & 4 fit in)

2. 13

prime
(only 1 & 13 fit in)

3. 16

composite
(1, 2, 4, 8 & 16 fit in)

4. 17

prime
(only 1 & 17 fit in)

5. 25

composite
(1, 5 & 25 fit in)

6. 19

prime
(only 1 & 19 fit in)

7. 20

composite
(1, 2, 4, 5, 10 & 20 fit in)

8. 21

composite
(1, 3, 7 & 21 fit in)

Now re-read the title I KNOW statement; decide whether you have 'Achieved' or need to revisit.

I KNOW every Times Table quick fact up to the 12x Table... The Ultimate Times Table Challenge!

Here is a list of all 2x to 12x Tables quick facts WITHOUT answers.

We are preparing for the '**Unit 4 Ultimate Challenge!**' on the next page.
The aim is to answer all 2x to 12x Tables quick facts instantly.

Use this page regularly to see how many you can answer instantly.

- Say them backwards sometimes so you get used to hearing the Times Table both ways
- Start in a different place each visit so you don't get used to the order

Try

3 x 11	3 x 12	12 x 9	3 x 9
3 x 3	4 x 12	11 x 12	12 x 2
4 x 2	7 x 7	11 x 11	10 x 6
8 x 7	10 x 10	3 x 2	5 x 5
5 x 3	4 x 11	8 x 6	10 x 7
6 x 5	7 x 4	4 x 3	2 x 8
12 x 5	5 x 7	11 x 10	11 x 9
9 x 9	9 x 2	8 x 3	3 x 6
6 x 6	11 x 7	6 x 7	12 x 8
3 x 7	12 x 7	5 x 10	5 x 8
4 x 4	11 x 8	10 x 9	12 x 10
5 x 4	2 x 7	8 x 8	10 x 8
9 x 6	9 x 8	5 x 9	6 x 4
2 x 11	12 x 6	3 x 10	8 x 4
10 x 4	11 x 6	6 x 2	11 x 5
2 x 2	9 x 7	10 x 2	4 x 9
	2 x 5	12 x 12	

Test

Here is a list of all 2x to 12x Tables quick facts WITH answers.
Drum roll please: The 'Ultimate Times Table Challenge!'

Hooray! So you think YOU KNOW every Times Table quick fact up to the 12x Table?
Congratulations! This is a huge milestone!

The aim is to answer all 2x to 12x Tables facts instantly.
How: The helper and the student sit opposite each other. The helper asks the Times Tables below;
the student answers as fast as possible. The student does NOT look at this page.

3 x 11 = 33	3 x 12 = 36	12 x 9 = 108	3 x 9 = 27
3 x 3 = 9	4 x 12 = 48	11 x 12 = 132	12 x 2 = 24
4 x 2 = 8	7 x 7 = 49	11 x 11 = 121	10 x 6 = 60
8 x 7 = 56	10 x 10 = 100	3 x 2 = 6	5 x 5 = 25
5 x 3 = 15	4 x 11 = 44	8 x 6 = 48	10 x 7 = 70
6 x 5 = 30	7 x 4 = 28	4 x 3 = 12	2 x 8 = 16
12 x 5 = 60	5 x 7 = 35	11 x 10 = 110	11 x 9 = 99
9 x 9 = 81	9 x 2 = 18	8 x 3 = 24	3 x 6 = 18
6 x 6 = 36	11 x 7 = 77	6 x 7 = 42	12 x 8 = 96
3 x 7 = 21	12 x 7 = 84	5 x 10 = 50	5 x 8 = 40
4 x 4 = 16	11 x 8 = 88	10 x 9 = 90	12 x 10 = 120
5 x 4 = 20	2 x 7 = 14	8 x 8 = 64	10 x 8 = 80
9 x 6 = 54	9 x 8 = 72	5 x 9 = 45	6 x 4 = 24
2 x 11 = 22	12 x 6 = 72	3 x 10 = 30	8 x 4 = 32
10 x 4 = 40	11 x 6 = 66	6 x 2 = 12	11 x 5 = 55
2 x 2 = 4	9 x 7 = 63	10 x 2 = 20	4 x 9 = 36
	2 x 5 = 10	12 x 12 = 144	

How did you do? Are you 100% there?
Are there a few stubborn ones which you struggle to remember? If so, make flashcards of them.
You are amazing! Remember to revisit this page at least once a week - don't let them get rusty.

FLASH FACTS ☆™

This final chapter is fun and light relief!

What is FLASH FACTS ☆™ ?

It's a list of **rapid-response questions, answers and quick activities** confirming the student's understanding of essential must-know facts; a randomly ordered collection of concepts, skills and tricks from Units 1, 2, 3 and 4.

☆   It is fun to use as a lesson warm up, mid lesson break, or close of lesson, usually in 10-15 minute bursts.

☆   It can be used in isolation when there is a spare ten minute opportunity, e.g. whilst waiting at the dentist.

☆   An occasional full hour session on **FLASH FACTS** ☆™ is sometimes appropriate once progress through the Units is being made.

Important game rule

It is designed to be a 'no pressure' activity to celebrate the things the student DOES know! It is fine to say, "no idea!" or "don't know!" as an answer.

Before you start the **FLASH FACTS** ☆™ themselves, **DO READ** all the 'guidance for use' information provided next.

Don't skip it... it's important.

Guidance for Use

FLASH FACTS ☆™ can be used in 3 different ways:

Use right from the very start of working on the four Units: either picking out some 'easier' questions, or asking them all with the rule that it is ok to say, "no idea!" or "don't know!" as an answer: in this case you can quickly provide the answer if it is a quickie, therefore beginning to sow the seeds and 'drip feed' correct responses, which will then be expanded upon once work gets going on the Units.

or

Use mid-way through working on the four Units: start to use when *some* of the work has been covered, with the rule that it is ok to say, "no idea!" or "don't know!" as an answer. This means the student at least gets used to hearing the 'unknown' questions and vocabulary; you can decide whether to discuss the answer or skill if it is a quickie, or leave it until later to address.

or

Use after all four Units are completed: wait until everything is covered and all the 'I CAN / I KNOW' statements are achieved so the student has the necessary skills, knowledge and understanding to provide all the answers. It can be used for months after completion of the programme to keep skills live and responses hot.

To clarify further

In response to a 'no idea!' or 'don't know!' you can:

✓ Leave it and wait until the student has covered it in the main work Units

✓ Quickly provide the answer and ask the student to 'put it in your brain for later!' and maybe s/he will recall it next time

✓ If a longer answer or explanation is needed, stop and ask each other if you want to stop and discuss it more fully right then – go with the moment

✓ Quickly make a note or mark the paper to highlight a 'don't know it' question / concept / skill, and if appropriate, use it as a starting teaching focus for the following session

Guidance for Use

How to do it

Helper and student sit **opposite each other** across a table. Helper asks; student responds. Have **scrap paper** positioned in the middle so both student and helper can reach it, as some of the activities need a quick sketch or a written calculation; of course it also helps sometimes to quickly write something down as a visual aid.

A bag of coloured tokens worth different amounts (e.g. red 50 points, blue 20, white 10) makes it exciting… have a few 'buttons' in there too, with a value of 1,000! A token is pulled out and given for every correct answer. The helper has a random handful of tokens (unknown value) in a pot at the beginning, and both **compare the final scores** at the end. **Working out** the final score is useful mental addition; of course, the helper usually loses!

✓   **Answers, and sometimes tips, are provided** for ease of quick reference for the helper.

✓   **Start in a different place sometimes.** Keep it light and fast paced; there is a variety of mental, written, drawing and 'doing' questions and activities.

✓   **Use as** a lesson warm up, a 'mid-lesson break or change of focus', or an end of lesson reward. Regularly revisiting **FLASH FACTS**☆™ keeps them hot.

✓   Devise a **simple system** (colours or ticks) to note which skills are strong / weak.

✓   Discuss the 'rule' that it is always ok for the student to be able to say, "no idea!" or "don't know!" and decide whether it is a 'quickie' to discuss / address there and then, or whether to leave it for a future main teaching session.

✓   Make a **game** of it and **laugh** a lot; tokens for answers makes it fun. Load on praise and show amazement at how well the student is doing to answer a particular question. Make fun of the 'silly' questions they got wrong or didn't know the answer to, telling them, "We'll get you next time!"

✓   The **Unit number is shown** at the start of each question in case you want to ask questions from one Unit only, or need to refer back to the teaching Unit at any time.

Great results from using FLASH FACTS ☆™ :

This is a great opportunity to quickly revisit 'learned concepts' and hop from one skill to another with only one little question or activity for each concept; it keeps responses sharp.

You can cover a lot of ground using this 'hopping around' approach; the student stays attentive for a long time as each question requires a different approach, activity and focus.

FLASH FACTS ☆™

③ What does the D mean in 2D and 3D? *dimensions / directions*

③ How many sides on a: quadrilateral? *4*   pentagon? *5*   hexagon? *6*

① What is: double 23? *46*   double 42? *84*   double 34? *68*

③ A square has a side measuring 5 cm (helper 'draw' in air), what is its perimeter? *20 cm*

③ How many degrees in a quarter turn? *90°*

④ Do you have a nightmare Times Table fact you always get caught out on? Tell me…

② Complete the fraction rap: "Whatever you do to the top…" *"you have to do to the bottom…"*

① What number comes to mind if I say the prefix: dec? *10*   cent? *100*   kilo? *1,000*

③ Stand up and face a point in the room, then turn:
quarter turn clockwise / half turn clockwise / three-quarter turn clockwise

① How do we find a quarter of a number? *we can half it, then half it again*

② Which is bigger, a tenth or a hundredth? *a tenth is bigger*
(think of a bar of chocolate cut up into tenths, and another cut up into hundredths)

③ Why are the first four 9x Table facts (9, 18, 27, 36) helpful when looking at degrees in a circle? *because if you plop a zero on the end of the answers you have the 90, 180, 270 and 360 degrees (quarter, half, three-quarters & full turn)*

③ How many degrees in: acute angles? *less than 90°*   right angles? *90°*
    obtuse angles? *more than 90° but less than 180°*
    straight angles? *180°*   reflex angles? *more than 180°*

① Show me the knuckles trick we use to see how many days there are in each month of the year.

④ Why is it easy to multiply any whole number by 10? *just place a zero on the end*

① How many: mL in a L? *1,000 mL*   g in a kg? *1,000 g*   m in a km? *1,000 m*

② What fraction alarm bell rings in your head when you see:
25%? *quarter*   0.25? *quarter*   50%? *half*

① What is another term / word for 'average'? *mean*

① Say your 15x Table until you get to 75: *15, 30, 45, 60, 75*

④ Define 'square number' and give an example: *when you multiply a number by itself, the answer is a square number, e.g. 6 x 6 is 36* (36 is the square number)

④ Define 'square root' and give an example: *a number that when multiplied by itself gives a square number, e.g. 4 x 4 is 16* (4 is the square root of 16)

③ Construct a square on paper (not to scale; just label dimensions) with a perimeter of:
20 cm *each side 5 cm*   12 cm *each side 3 cm*   16 cm *each side 4 cm*

① Why might a hiding finger be useful when dealing with numbers which end in zero(s)?
it can make working with large numbers easier, e.g. 40 x 3: hide the zero so 4 x 3 = 12, then put the zero back on the end to make 120

② Give me an example of a mixed number: *e.g.* $1\frac{2}{5}$

④ Tell me about a Times Table 'trick' you know.

③ When talking about circles, what is: circumference? *distance around a circle*
radius? *distance from centre to 'side' of a circle*
diameter? *distance from one 'side' of a circle to the other, passing through the centre*

③ What is the formula for finding the area of a square or rectangle? *length x height*

③ What is the formula for finding the area of a triangle? $\frac{1}{2}$ *base x height*

① Using column multiplication, on paper, multiply 46 by 3; check your answer with a calculator: *138*

④ How many Times Tables quick facts can you answer in 60 seconds? Let's go...
(choose any currently working on, or select from page 244)

③ How are a rhombus and a kite different? *a rhombus has 4 sides all the same length; a kite has 2 equal shorter sides and 2 equal longer sides*

②   How would you write the fraction 'half' as a decimal? **0.5** or **0.50**

③   Draw a circle, and show: the radius **centre to 'side'**
                                 the diameter **'side' to other 'side' passing through centre**

①   When we use the term 'difference', which operation is this? **subtraction**

①   When we use the term 'product', which operation is this? **multiplication**

③   What number jumps to mind if I say the prefix: tri? **3**   quad? **4**   pent? **5**   hex? **6**

③   What is the formula for finding the volume of a cube or a cuboid?
       length x height x width

② ✎   What is the 2 worth in 75.26? **2 tenths**

③   What do we call an angle that is: less than 90°? **acute**   exactly 90°? **right**
       more than 90° but less than 180°? **obtuse**   exactly 180°? **straight**   more than 180°? **reflex**

①   How many mL in: 2 L? **2,000** mL   4 L? **4,000** mL   5 L? **5,000** mL

②   Name the 3 types of fractions: **proper, improper** and **mixed number**

②   Draw a rectangle and divide it as accurately as you can into halves, then quarters, then eighths.

①   Count up in twenties until you reach 100: **20, 40, 60, 80, 100**

③   How many sides on a: trapezium? **4**   parallelogram? **4**   kite? **4**   rhombus? **4**

③   When referring to 2D shapes such as pentagons, hexagons and octagons, what does:
       'regular' mean? **same length sides and same internal angles**
       'irregular' mean? **some different length sides and different internal angles**

③   Draw an irregular: pentagon (5 sides) / hexagon (6 sides) / octagon (8 sides) / decagon (10 sides)
       (if age appropriate, make them into creatures, cars or monsters)

④ ✎   Multiply this whole number by 10:   43 **430**   875 **8,750**   78 **780**   236 **2,360**
       (to multiply a whole number by 10, place a zero on the end)

④   What is the name given to the tiny ^{2} or tiny ^{3} placed after a number? **an exponent**

①   Name 2 months which have only 30 days: **April, June, September, November**

③   What prefix jumps to mind for the number: 4? **quad**   5? **pent**   6? **hex**   8? **oct**   10? **dec**

③   Use a protractor to construct an angle that is: acute **less than 90°**   right **exactly 90°**
       obtuse **more than 90° but less than 180°**   straight **exactly 180°**

①   'Show' me a: mm / cm / dm / m using your fingers, arms or ruler

③   If a rectangle measures (helper 'draw' in air) 4 cm, 10 cm, 4 cm, 10 cm, what is its perimeter?
       28 cm *(distance around the shape, so add all 4 numbers)*

①   How many metres in: 3 km? **3,000 m**   4 km? **4,000 m**   6 km? **6,000 m**

②   What is one third of: 9? **3**   18? **6**   30? **10**   *(count up in threes)*

②   What is one fifth of: 20? **4**   45? **9**   25? **5**   *(count up in fives)*

③   How many sides on: an octagon? **8**   a decagon? **10**

③   If a pair of 2D shapes are congruent, what does this mean? **they are identical in shape and size**

②   Count up in 25s until you get to 100: **25, 50, 75, 100**

④   Count up in: 2s / 3s / 4s / 5s / 6s / 7s / 8s / 9s / 10s / 11s / 12s until I say stop

①   What is the remainder if I divide: 32 by 5? **2**   19 by 4? **3**   13 by 3? **1**

④   Count backwards in: 2s / 5s / 10s, starting at 40

③   Stand up and face a point in the room, then turn:
       90° anti-clockwise / 180° anti-clockwise / 270° anti-clockwise

④   Test me on five tricky Times Tables quick facts and correct me if my answers are incorrect.

③   Draw your best: cube / triangular prism / cuboid

① ✎   What is the 8 worth in 86.2? **8 tens (80)**

①   How many 5s in 100? **20**

①   How many 20s in 100? **5**

②   How many 25s in 100? **4**

①   Using short division, on paper, divide 252 by 3; check your answer with a calculator: **84**

③   If a rectangle measures (helper 'draw' in air) 3 cm, 10 cm, 3 cm, 10 cm, what is its area? **30 cm²**
       (length x height)

②   What is 50% of: 20? **10**   30? **15**   60? **30**   46? **23**   *(find half)*

②   What is 25% of: 80? **20**   40? **10**   60? **15**   200? **50**   *(find a quarter)*

②   What is 75% of: 80? **60**   40? **30**   60? **45**   200? **150**
       (find three-quarters by finding one quarter first, then x3)

①   How many years in a: decade? **10 years**   century? **100 years**

②   How many mL in half a L? **500 mL**

①   What is another term / word for 'mean'? **average**

④   Show me and explain how the 9x Table 'tucking in fingers trick' works.

③   How many faces on a: cuboid? **6**   triangular prism? **5**   cube? **6**
                                      square pyramid? **5**   triangular pyramid? **4**   rectangular pyramid? **5**

①   What is the product of: 4 and 5? **20**   6 and 10? **60**   70 and 2? **140**   3 and 5? **15**

①   What is double: 60? **120**   32? **64**   55? **110**   700? **1,400**

②   How would you write the fraction 'quarter' as a decimal? **0.25**

③   What number jumps to mind if I say the prefix: sept or hept? **7**   oct? **8**   non? **9**   dec? **10**

①   How many 15s are there in: 60? **4**   45? **3**   30? **2**

②   Draw a rectangle and divide it as accurately as you can into thirds, then sixths.

①   What is the difference between: 6 and 10? **4**   3 and 9? **6**   70 and 40? **30**   123 and 125? **2**

② ✎   What is the 5 worth in 123.75? **5 hundredths**

①   Multiply: 4 by 90 **360**   900 by 3 **2,700**   400 by 5 **2,000**   300 by 7 **2,100**
                   (maybe use the hiding finger)

④   Use the 'pop-up-finger-method' to show me: e.g. 6x Table *(choose any currently working on)*

③   'Draw' this shape in the air: trapezium / parallelogram / kite / rhombus

①   What is half of: 40? **20**   30? **15**   600? **300**   70? **35**   82? **41**

② ✎   What is: 3.45 multiplied by 10? **34.5**   16.2 multiplied by 10? **162**
    17.53 multiplied by 10? **175.3** *(decimal point jumps one place to the right)*

①   Find a quarter of: 8 **2**   20 **5**   60 **15**   80 **20**   *(half it, then half it again)*

② ✎   What is the fraction $\frac{3}{50}$ as a decimal? **0.06**
    (multiply top and bottom by 2 to make it a fraction out of 100 first)

③   Tell me about the sides of a scalene triangle: *all 3 sides are different lengths*

③   Tell me about the sides of an isosceles triangle: *2 sides same length, 1 different length*

③   Tell me about the sides of an equilateral triangle: *all sides same length*

③   Draw, name and label 6 different quadrilaterals:
    square, rectangle, trapezium, rhombus, kite, parallelogram

①   Mentally divide 30 by: 3 **10**   5 **6**   6 **5**   10 **3**   *(count up in the divisor)*

②   How many mL in quarter of a L? **250 mL**

② ✎   Express: $4\frac{1}{4}$ as a decimal **4.25** $2\frac{1}{2}$ as a decimal **2.5** $5\frac{3}{4}$ as a decimal **5.75**

①   Write one million in digits: **1,000,000**

②   If a calculator reads 4.5 as the answer to a money problem, how would you say the answer? **four pounds and 50 pence** (the 5 is in the tenths column)

②   If a calculator reads 6.05 as the answer to a money problem, how would you say the answer? **six pounds and 5 pence** (the 5 is in the hundredths column)

④   What is: 3 squared? **9** (3 x 3)   4 squared? **16** (4 x 4)   5 squared? **25** (5 x 5)   6 squared? **36** (6 x 6)

② ✎   Change this improper fraction into a mixed number: $\frac{7}{4}$ **$1\frac{3}{4}$** $\frac{14}{5}$ **$2\frac{4}{5}$** $\frac{7}{2}$ **$3\frac{1}{2}$** $\frac{17}{3}$ **$5\frac{2}{3}$**

③   Using 2 rulers, show me: parallel **like train tracks, the 2 lines never meet**
intersecting **meeting or crossing**
perpendicular **meeting or crossing to make a right angle; an 'L' shape is formed**

①   How many g in: 2 kg? **2,000 g**   3 kg? **3,000 g**   6 kg? **6,000 g**

③   What do we call the distance from the centre of a circle to the 'side'? **radius**

③   What do we call the distance from one 'side' of a circle to the other 'side', passing through the centre? **diameter**

③   What do we call the distance around a circle? **circumference** (notice how it starts with 'circ')

② ✎   What is: 3.45 divided by 10? **0.345**   16.2 divided by 10? **1.62**
    17.53 divided by 10? **1.753**   (decimal point jumps one place to the left)

②   How would you write the fraction 'three-quarters' as a decimal? **0.75**

③   How many vertices on a: cuboid? **8**   triangular prism? **6**   square pyramid? **5**
    triangular pyramid? **4**   rectangular pyramid? **5**   cube? **8**

③   Use a protractor to construct an angle that is: 50° / 35° / 105° / 150° / 170°

① ✎   What is the 5 worth in 75,124? **5 thousands** (5,000)

③   If a square has a side measuring 5 cm, (helper 'draw' a square in air) what is its area? **25 cm²** (length x height)

② ✎   What is: 5.34 multiplied by 100? **534**   25.66 multiplied by 100? **2,566**
    4.592 multiplied by 100? **459.2**   (decimal point jumps two places to the right)

②   Using a calculator, change this fraction into a decimal: $\frac{3}{12}$ **0.25** $\frac{3}{30}$ **0.1** $\frac{2}{8}$ **0.25**
(divide the top number by the bottom)

①   Name 3 important things to remember when you do a worded problem: **e.g. highlight important information, make the four operations sausage and re-read when you reach the answer**

③   Draw your best: square pyramid / rectangular pyramid / triangular pyramid

③   How many degrees in a half turn? **180°**

④   How many: 5s in 30? **6**   4s in 12? **3**   6s in 36? **6**

③   Name three transformations, and describe them:
   translation: slide   ***rotation: turn***   ***reflection: flip***

②   Give me an equivalent fraction to a half: **e.g.** $\frac{5}{10}$ *or* $\frac{4}{8}$ *(any where denominator 2x numerator)*

②   Give me an equivalent fraction to a quarter: **e.g.** $\frac{2}{8}$ *or* $\frac{3}{12}$ *(any where denominator 4x numerator)*

④   How long will it take you to write out the products of, e.g. 4x Table? Let's go...

②   How many g in three-quarters of a kg? **750 g**

①   Name 2 months which have 31 days: ***January, March, May, July, August, October, December***

④   Working mentally, what is: 55 divided by 11? **5**   16 divided by 4? **4**   25 divided by 5? **5**
     (count up in the divisor)

③   How many edges on a: cuboid? **12**   triangular prism? **9**   cube? **12**
     square pyramid? **8**   triangular pyramid? **6**   rectangular pyramid? **8**

②   Draw a circle and divide it as accurately as you can into thirds, then sixths.

①   Tell me some words or terms which mean: add / subtract / multiply / divide

②   What fraction alarm bell rings in your head when you see:
   0.5? **half**   75%? ***three-quarters***   0.75? ***three-quarters***

③   Measure this angle with a protractor: *(helper construct an angle)*

④   Is 24 / 50 / 49 / 36 / 25 / 13 a square number and why? ***49 is (7 x 7), 36 is (6 x 6), 25 is (5 x 5)***
(24, 50, 13 are not as they are not the answer to a number multiplied by itself)

②   Find three-quarters of: 8 **6**   20 **15**   60 **45**   800 **600**   *(find one quarter first, then x3)*

①   Use a calculator to find the answer to: difference between 172 and 85 **87**
   sum of 567 and 42 **609**   product of 77 and 4 **308**

② ✎   Simplify this fraction: $\frac{3}{9}$ *$\frac{1}{3}$* $\frac{8}{16}$ *$\frac{1}{2}$* $\frac{6}{9}$ *$\frac{2}{3}$* $\frac{15}{25}$ *$\frac{3}{5}$*

①   How do we find the mean (average) of a set of numbers?
   add them up, then divide by how many there are

①   How do we find the median of a set of numbers?
   put them in numerical order and find the middle number

①   How do we find the mode of a set of numbers? ***it is the number which appears most often***

①   What is: 20 x 4? **80**   15 x 3? **45**

③   Construct a rectangle on paper (not to scale; just label dimensions) with a perimeter of:
10 cm **e.g. 1, 4, 1, 4 cm** or **2, 3, 2, 3 cm** (both total 10 cm)

12 cm **e.g. 1, 5, 1, 5 cm** or **2, 4, 2, 4 cm** (both total 12 cm)

② ✎   What is the fraction $\frac{2}{20}$ as a percentage? **10%**
(multiply top and bottom by 5 to make it a fraction out of 100 first)

④   Tell me (or write) some square numbers: **e.g. 4, 9, 16, 25, 36, 49, 64, 81, 100, 121, 144**
(a square number is the product when we multiply a number by itself)

② ✎   What is: 7.23 to the nearest tenth? **7.2**   23.56 to the nearest tenth? **23.6**
2.55 to the nearest tenth? **2.6**

② ✎   What is: 8.242 to the nearest hundredth? **8.24**   12.655 to the nearest hundredth? **12.66**
25.547 to the nearest hundredth? **25.55**

①   Using short division, on paper, find half of 9,632; check your answer with a calculator:
4,816 (9,632 divided by 2)

②   What is 25% of: 40 dogs? **10 dogs**   60 eggs? **15 eggs**   8 fish? **2 fish**   (find quarter)

②   What is 50% of: £44? **£22**   300 pencils? **150 pencils**   24 cats? **12 cats**   (find half)

②   What is 75% of: 40 dogs? **30 dogs**   600 eggs? **450 eggs**   80 fish? **60 fish**
(find three-quarters, so find one quarter first, then x3)

②   Give me an example of a proper fraction: **e.g.** $\frac{2}{5}$ (any where numerator is smaller than denominator)

②   How many 25s are there in: 75? **3**   50? **2**   100? **4**

② ✎   Change this mixed number into an improper fraction: $1\frac{3}{4}$ **$\frac{7}{4}$** $4\frac{3}{5}$ **$\frac{23}{5}$** $2\frac{4}{9}$ **$\frac{22}{9}$** $3\frac{4}{5}$ **$\frac{19}{5}$**

③   How many degrees in a three-quarter turn? **270°**

① ✎   What is the mean (average) of these darts scores: 9, 8, 9, 11, 13? **10**   5, 3, 6, 5, 7, 4? **5**
(add them up and divide by how many there are)

①   What is half of: 7? **3.5**   9? **4.5**   13? **6.5**   15? **7.5**
                         300? **150**   150? **75**   180? **90**

④   What is: 7 x 8? **56**   6 x 7? **42**   8 x 6? **48**   11 x 12? **132**   11 x 11? **121**   9 x 12? **108**

② ✎   A store offers 25% off its books: if books are usually £16, what is their discounted price? **£12**
if books are usually £40, what is their discounted price? **£30**
(find a quarter (25%), and take it off the price)

②   Draw a circle and divide it as accurately as you can into fifths, then tenths.

② ✎ Add these fractions: $\frac{2}{5} + \frac{1}{5} = \frac{3}{5}$ $\frac{3}{6} + \frac{2}{6} = \frac{5}{6}$ $\frac{6}{10} + \frac{3}{10} = \frac{9}{10}$

④ If we are asked to 'cube' a number, what does this mean? **it is when we take a number and multiply it by itself, then multiply it by itself again** (e.g. 5 cubed (5^3) is 5 x 5 x 5)

③ How many degrees in a full circle turn? **360°**

② ✎ What is $\frac{7}{8} - \frac{2}{8}$? $\frac{5}{8}$

④ What is a prime number? **a number only divisible by one and itself** (no other number will fit in)

① ✎ What is the median of these darts scores: 9, 8, 9, 11, 13? **9**   5, 3, 6, 5, 7, 4? **5**
(put in numerical order and find middle)

③ Construct a square on paper (not to scale; just label dimensions) with an area of:
36 cm² **6 cm x 6 cm**   16 cm² **4 cm x 4 cm**   4cm² **2 cm x 2 cm**

① What is the sum of: 7 and 2? **9**   50 and 30? **80**   2 and 3? **5**

② What place value is: the first column after a decimal point? **tenths**
the second column after a decimal point? **hundredths**

③ If the radius of a circle is 4 cm, what is the diameter? **8 cm**

③ Draw your best: cone / cylinder / sphere

① What is a quarter of: 200? **50**   800? **200**   1,000? **250**   120? **30**
(half it, then half it again)

② Give me an example of an improper fraction: **e.g.** $\frac{7}{5}$
(any where numerator is larger than denominator)

④ What is the square root of: 25? **5**   49? **7**   81? **9**   36? **6**

① Using a calculator, divide 6 into 552: **92** (552 ÷ 6... not 6 ÷ 552)

③ If a triangle's sides are all different lengths, what is it called? **scalene triangle**

③ If a triangle has 2 sides the same length and 1 side a different length, what is it called?
isosceles triangle

③ If a triangle has sides all the same length, what is it called? **equilateral triangle**

② How many grams in $1\frac{3}{4}$ kg? **1,750 g**

② What is two thirds of: 9? **6**   12? **8**   30? **20**   24? **16**
(find one third first by counting up in threes, then x2)

④ Can you tell me some prime numbers? **e.g. 2, 3, 5, 7, 11, 13, 17, 19, 23, 29**

④ What do we call a number that is NOT a prime number? **a composite number**

①   Write this 'big' number in digits: one hundred and thirty thousand **130,000**
       four million **4,000,000**   twenty three thousand and four **23,004**

④   What is: 2 cubed? **8** *(2 x 2 x 2)*   3 cubed? **27** *(3 x 3 x 3)*   4 cubed? **64** *(4 x 4 x 4)*

① ✎   What is the mode of these darts scores: 9, 8, 9, 11, 13? **9**   5, 3, 6, 5, 7, 4? **5**
                                         (the number occurring most often)

①   Tell me some multiples of 4: *e.g. 4, 8, 12, 16, 20, 24 etc.*

② ✎   What is: 45.32 divided by 100? **0.4532**   124.5 divided by 100? **1.245**
             143.28 divided by 100? **1.4328**   *(decimal point jumps two places to the left)*

③   Construct a rectangle on paper (not to scale; just label dimensions) with an area of:
       8 cm² *e.g. 1 x 8 or 2 x 4 cm*   15 cm² *e.g. 1 x 15 or 3 x 5 cm*
       12 cm² *e.g. 1 x 12 or 2 x 6 or 3 x 4 cm*   *(length x height)*

③   What do the internal angles of all triangles add up to? **180°**

③   What do the internal angles of all quadrilaterals add up to? **360°**

①   How many days in: June? **30**   March? **31**   December? **31**   February? **28** or **29**

③   If the diameter of a circle is 6 cm, what is the radius? **3 cm**

①   Using column multiplication, on paper, multiply 456 by 23; check your answer with a calculator:
       10,488

②   What is three fifths of: 20? **12**   30? **18**   25? **15**   50? **30**   55? **33**
       (find one fifth first by counting up in fives, then x3)

④   Is 15 a prime number? **no** *(3 and 5 will fit into it as well as 1 and 15)*

③   Draw your best: isosceles triangle / scalene triangle / equilateral triangle